MARVELS
OF
MODERN ELECTRONICS
A SURVEY

MARVELS
OF
MODERN ELECTRONICS
A SURVEY

Barry M. Lunt, Ph.D.

Full Professor of Information Technology
Brigham Young University
Provo, Utah

Dover Publications, Inc.
Mineola, New York

Bibliographical Note

Marvels of Modern Electronics: A Survey is a new work, first published by Dover Publications, Inc., in 2013.

International Standard Book Number
ISBN-13: 978-0-486-49838-6
ISBN-10: 0-486-49838-7

Manufactured in the United States by Courier Corporation
49838701
www.doverpublications.com

To my loving and amazing wife, Cathy;
to my wonderful children Greg, Cindy, (Doug) and Brad;
and to the many who have been excited with me
to see this book published.

CONTENTS

INTRODUCTION

Up until about 1800, we knew very little about the true nature of electricity, and no practical devices existed which used electricity. Its several manifestations were lab curiosities and interesting displays at sideshows and physics societies. It is by no means an overstatement to say that since 1800, advances in our knowledge and application of electricity have been one of the greatest marvels of modern times. The advances made have been unparalleled by any other field, by essentially any measure. This is not to disparage in any way the tremendous advances made in many fields over the past several decades, especially including the fields of biochemistry, nanotechnology, chemistry, manufacturing, and others. It is this author's desire to outline the amazing discoveries and inventions that have come about in electronics, and to describe some of the things electronics has enabled.

As in any field of development, there are names that quickly rise to the top, followed closely by other names near the top, and so on down to the many thousands of individuals, who though historically nameless, each made a contribution to the field. I wish to acknowledge all, but only the names of a few will have to suffice. All progress has come as a result of great people who built on the work of others, as expressed so aptly by Sir Isaac Newton: "If I have been able to see further, it was only because I stood on the shoulders of giants."

And so it has been with electricity. All we have learned has come one little piece at a time, sometimes by serendipity, other times by great, focused effort. Step by tiny step, we learned more about the true nature of electricity. Step by tiny step we learned how to apply this knowledge to create useful products, which would not have been possible without this knowledge. And as more products were created, they further enabled more learning about electricity and the new products themselves. An example from the field of transportation illustrates this point.

Henry Ford's process of mass production enabled the manufacturing of much less expensive automobiles, which in turn further enabled transporting goods, which in turn brought less expensive parts and materials, which in turn further reduced the cost of mass production. As each step improved, it further enabled closely related fields, which caused the improvements to loop back around and increase the improvements in mass production.

Over the decades, many individuals and organizations have published lists of the greatest inventions of mankind. I have seen many of these lists, and it is difficult to argue that each of the inventions listed is indeed very significant. Given my background in electronics, I always scan these lists looking for how many of them are directly a result of electricity, and in all these lists, one thing I have seen in common is that about one-third of them are.

These lists commonly include things like the wheel and the printing press, things which are pretty easy to understand, and things which have had (and continue to have) a huge impact on our daily lives. But these lists also usually include things like electricity, the telephone, radio, television, computers, the transistor, the integrated circuit, the Internet—all things that also have had (and continue to have) a huge impact on our daily lives, but things which we generally do not understand very readily.

I was born an engineer; from the earliest I can remember, I always looked at things and asked myself how they worked. I asked my mother how a light bulb worked; I was fascinated how music could come out of a box that seemed to have nothing going into it (a radio); I could not fathom how a moving image of something could be transmitted through thin air (television). And when I eventually learned how radio and television work, my fascination grew even more. Every time I have a chance to learn how something works, I love learning about it.

I once had the opportunity to take a tour of a Boeing manufacturing facility in Seattle, Washington. It was absolutely fascinating to see how an airplane is put together. Now, every time I ride in a plane, I think of the walls of that plane that are protecting me, how the wings are built and how they function, how the engines work, and how all the systems of the plane come together to provide such an amazing system. This same kind of experience happens every time I learn how something works—I never look at it the same, and I'm filled with amazement and gratitude that there are people and companies out there that can see a need, and design and build a product that meets that need.

I assume there are a lot of people out there who also enjoy learning about how things work; I know I have met a lot of them in my lifetime, and I have read many of their books and enjoyed them very much. Over the past 30 years, I have been continually fascinated by the major advances in our civilization that have been enabled by electricity and the inventions that depend on it. If you were to take away electricity, society as we know it today would totally disappear. We would be back to only steam or diesel-powered automobiles; there would be no airplanes; no telephones or cell phones; no Internet; no computers or computer-related equipment of any kind; no electronic entertainment industry (movies, video games, music); no stop lights; no public-address systems (sound amplification); no widely affordable books; no electric motors (which also means no refrigerators, air conditioning, fans, freezers, manufacturing equipment, pumps, etc.); no modern medical equipment (at least nothing that must be plugged in, which is most of it); and no lights (except for fires). Essentially, we would be back to about how things were in the 1850s, with the exception of a few advances that did not involve electricity in any way.

Electricity, and all the products that depend on it, have impacted our life in truly amazing ways. Today, most of those products are based on electronics—vacuum tubes, transistors, and integrated circuits. So if you've ever wanted to understand how those products function, this book is for you. It covers topics I have taught at the college level for many years, topics I have worked hard to help students understand. And I hope this book conveys my fascination with how amazing electronic technology really is.

This book takes the reader through the rudimentary history of electricity, up through the creation of electrical and electronic products, and describes several of these products and their impact on society and on further advances in electronics. This piece of our history and the changes it has wrought are truly some of the greatest marvels of our modern times.

CHAPTER 1
ELECTRICITY: A RUDIMENTARY HISTORY

Static Electricity

The earliest record we have of electricity and its properties comes from the classical Greek civilization roughly around 600 B.C.E. They noticed that rubbing fur against a piece of amber ("elektron" in Greek) caused the amber to attract bits of dust, straw, or feathers. They did not come to learn a way to explain this behavior, nor did they create any useful products from the study of this phenomenon.

For approximately 2,000 years, this was about as much as we knew about what came to be called "electricity." This particular form of electricity was termed static electricity; "static" means not changing or moving. Experimenters were able to build up a substantial charge of static electricity by rubbing various non-conductive materials against each other. They noticed that some materials behaved differently than others. For example, a piece of fur rubbed against amber would *attract* a piece of silk rubbed against glass, but two pieces of fur, each rubbed against amber, would *repel* each other.

Experimenters in the early 1600s created experimental devices using balls of sulfur, amber, and other materials. By rotating these materials against silk or fur, they were able to build up substantial charge, which could readily create a spark (or shock the experimenter!)

Having no other mental model readily available to understand this phenomenon, electricity was thought of as a fluid. Accordingly, in order to store this fluid, it was natural to think of storing it in a jar. Glass jars were readily available, but how to get one to hold this fluid? It was known that glass would not conduct this fluid, but that metals would. The answer was to put a lining of metal on the inside of the jar, and to connect this metal to the fluid. The electricity should then flow inside the jar and be retained there. However, when this was attempted, no electricity seemed to be stored in the jar. At the University of Leyden in the Netherlands (about 1746), the jar was also coated with metal on the outside. This device, later known as the

1

Leyden jar, was capable of storing a substantial amount of this electric fluid, as evidenced by the sparks or shocks it could produce. Today we know that a Leyden jar was an early type of capacitor, a device which can hold small amounts of electric charge. A common application for a capacitor today is to hold the charge from a battery so it can be quickly released in a flash for a camera.

It should be appreciated that in these early days of experimentation, no devices were available to allow the experimenters to quantify their results and thereby be able to perform truly scientific experiments. One couldn't measure the amount of static electricity being generated or stored in a Leyden jar, nor how energetic the spark was when the electricity was discharged. One could only say that a certain discharge was louder, or the spark was longer, or the shock received by the experimenter was greater. But such were the times, and such was the determination of the experimenters that this did not dissuade them from their studies.

A classic example of this is the kite-flying experiment of Benjamin Franklin. The question he sought to answer: Was lightning the same thing as this static electricity which people had been generating, storing, and studying? To answer this question, he equipped himself with the standard tools of the time, which included a Leyden jar, a kite, and a skeleton-type key, which he tied near the bottom end of the string. The thought was that, on a cloudy day when lightning was striking, the kite would pick up some of this "fluid;" it would travel down the string, through the key, and then into the Leyden jar. His kite flew, the key supplied sparks (which he felt by letting them discharge to his wrist), and he was able to store some of this fluid in a Leyden jar. His experiment was a great success, but he could easily have been killed (as a few others were who repeated his experiment)! He later proved that lightning was indeed the same as static electricity, by using the filled Leyden jar to perform experiments already familiar to himself and other electrical experimenters.

Many worked on the idea of producing some kind of useful work with this electricity. The steam engine was already in use in many applications, but it was large, slow, heavy, and potentially very dangerous. If an electric motor could be made, it would be of great use in many applications. Over the years, several types of static electricity motors were made, but they proved far too weak to perform useful work. Static electricity was just too ephemeral, and it was always a challenge to keep it from arcing to places it should not go.

In the end, even today there are relatively few practical products that use static electricity. Great sideshow equipment exists, such as Jacob's ladders, plasma balls, Tesla coils, and Van de Graaff generators, but only the last has much in the way of practical uses, and then only in very specialized devices, such as particle accelerators. There is one type of home smoke detector that works using static electricity, and electrostatic precipitators have done much to help clean up smoke-stack emissions. Additionally, static electricity is used in photocopiers and sometimes in painting and coating. There are surely other niche applications of which the author is unaware, but compared to normal (or "dynamic") electricity, they are much less common.

A New Kind of Electricity

In the early 1770s, Luigi Galvani was experimenting with the muscles of frogs. Using electrostatic generators or Leyden jars to store electricity, he would apply some of this electric fluid to the muscles and they would twitch. He wisely concluded that there were electrical signals at work in a frog's body, causing the muscles to respond accordingly. Later work on human bodies gave similar results, which gave great insight into how the human body works.

On his workbench, there was a metal covering of brass. The dissecting instruments were made of steel. He noticed that, occasionally, there was a twitch of the muscles when he applied the cutting blade, leading him to conclude that there was an electrical charge on the blade. While many others accepted his theory, Alessandro Volta did not. Indeed, by further experimentation, Volta (about 1800) showed that dissimilar metals in a moist environment actually created electricity. Thus, it was the brass and steel, in the environment of the moist frog, that created the electricity. This was a keen insight, and proved very useful.

Volta went on to create several stacks (or "piles") of these electrical sources, which we now know as batteries, and did a great deal of experimentation with them. Each of Volta's cells were made of zinc and copper, separated by a space filled with a brine solution. By stacking them, he found he could create more electricity.

3

Again, lacking any volt meters (the Volt was later named after Volta) or measuring instruments of any kind, Volta was left to his own devices to determine how to conduct experiments with these batteries. Knowing that his body could sense electricity, it made sense to him to use it. Accordingly, his lab book records that he would touch various parts of his body with electrodes attached to different batteries. If the sensation was stronger with one battery than another, he would correctly conclude that the stronger sensation was due to more electricity. On one occasion, he recorded that when he touched a fresh wound with these electrodes, the sensation was particularly acute. It is amazing that in spite of years of experiments of this type, he did not die of electric shock!

The biggest difference between Volta's "batteries" (a term coined by Benjamin Franklin; others called them "Voltaic piles") and the electrostatic generators of the time was that the battery gave a constant current, rather than momentary discharges. It was thus the first practical source of electricity. Although these batteries did eventually discharge and could not be recharged, they were nevertheless much more practical than the electrostatic generators. Finally experimenters had a source of electricity that would remain constant for several hours or days, depending on their application. One of the first practical applications they would see was in the telegraph, starting in about 1850. Each telegraph transmitter/receiver required some source of electricity to operate. The simple opening and closing of an electric switch (the telegraph "key") generated the dots and dashes of the Morse code; batteries were the only source of electricity which could supply this need.

And Yet Another Kind (of Electricity)!

For centuries, scientists were fascinated by the properties of magnetism. The manifestations of this phenomenon were only available from lodestone, a naturally occurring magnetic material found in many places around the earth. It was common for physics professors to use lodestone magnets and compasses to demonstrate various properties of this force. It was also common for physics professors, after the widespread availability of batteries, to demonstrate various properties of electricity. This often included showing how electricity could be conducted through wire.

4

During one of these demonstrations, the physicist Hans Christian Oersted was busy demonstrating the heating of iron wire as electricity from a battery passed through it. He noticed that each time he connected the battery to the wire, the nearby magnetic compass moved. He worked for months experimenting with this newly discovered phenomenon, keeping it a secret, but was unable to explain the results. Eventually, he chose to publish his findings without explanation.

Soon after Oersted published his findings, the field began to expand rapidly. Andre-Marie Ampere showed that two wires conducting electricity exerted a magnetic force on each other. The insightful and careful experimentalist Michael Faraday showed that not only did electricity flowing in a wire create magnetism, but the reverse was true: a magnet moving near a wire caused electricity to flow in the wire! Soon after, he produced the first electric motor and the first electric generator. When asked what good they were (the electric motor and generator), he replied, "What good is any newborn baby?" Indeed, his motor and generator were initially of very little interest, since there were no applications! But the eventual impact of these inventions was staggering; today well over 95% of all electricity generated uses the principles he demonstrated with his electric generator, and nearly 100% of all

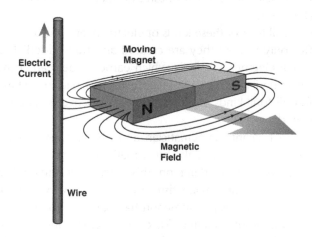

Figure 1-1:
A magnet moving near a wire (or any conductor) will produce a flow of electricity.

motors use the principles he demonstrated with his first electric motor. Even internal combustion engines, both gasoline and diesel, use electric motors as starter motors and use generators (commonly called alternators) to keep the batteries charged.

The invention of the electric generator and motor all occurred by about 1820. It would take several decades before the electrification of the United States was well underway (about 1890), but in localized applications, the electric generator began to supply energy to drive electric motors, which began to replace the steam engines common to the era. Today, in nearly all applications where electricity is available, electric motors provide the driving mechanical force. It is also true that if large quantities of electricity could be stored easily, automobiles would all be powered by electric motors, and the internal combustion engine would become a rarity.

And Yet Another (Type of Electricity)?

The end of the previous section leaves us with three kinds of electricity: static, battery-supplied, and generator-supplied. As has been mentioned, static electricity has historically been much less useful than the other two types, and this is still true today. But there is a fourth kind of electricity, a kind that doesn't normally seem much like electricity at all to us.

In reality, all four of these kinds of electricity are one and the same, and differ only in how they are created and transmitted. The question of what electricity really is, which earnestly begs an answer, will have to wait until a subsequent section, just as it had to wait in history until after all four kinds of electricity were discovered and began to be exploited.

In 1864 a brilliant mathematician and physicist (ever notice how physicists are nearly always also great mathematicians?) named James Clerk Maxwell was working on this very interesting relationship between electricity and magnetism. Since its discovery over 40 years earlier by Oersted, this phenomenon had basically gone unexplained but had not gone unexploited. Maxwell's breakthrough was to show mathematically how these two forces were interrelated, and in the process he showed that the speed of light was included in the relationship. This meant that these two forces, electricity and magnetism, were actually different manifestations of the same force, which we now call electromagnetism. It also meant that in a vacuum, these electromagnetic forces would travel at the speed of light. Maxwell also predicted that light itself would be discovered to be simply another manifestation of this same force. Although Maxwell's predictions seemed incredible at the time, the subsequent years soon showed his conclusions to be correct.

Following on Maxwell's work, another great experimenter named Heinrich Hertz was able to use a simple spark gap (like the spark plugs in an automobile engine) and some iron filings to show how the electromagnetic waves predicted by Maxwell indeed existed, and how they could be projected across a room and detected on the other side. This work was quickly followed by developments from Guglielmo Marconi, in which he extended the range of his ability to send and detect these very weak waves. Working in his lab, he successfully repeated Hertz's experiments, then found ways to extend the distance beyond his lab, beyond the estate where he worked, and beyond the confines of what he could easily put together for his experiments. Eventually he found his way up from Italy to France, where he successfully transmitted and received Morse code across the English Channel. In December of 1901, he culminated this work by successfully transmitting and receiving the letter "s" in Morse code across the Atlantic Ocean, from Scotland to Newfoundland.

This transatlantic accomplishment deserves an important side note. Marconi knew very well that these "radio" waves he was working with were electromagnetic waves, and that they traveled only in straight lines. He also knew very well that the Earth was round. It made very little sense to expect that he could send these waves all the way from Scotland to Newfoundland, following the curvature of the Earth! But regardless of the impossibility of this feat, he succeeded! Today, we know he succeeded because of the ionosphere, an electromagnetically charged layer of the atmosphere off which his radio waves bounced back to Earth. But no one knew of such a layer in 1901!

The development of the radio (then called the "wireless") helped solve the vexing problem of how to communicate with ships. Telegraph required the presence of a wire, over which the telegraphed signals traveled; obviously this would never work for ships. Marconi's developments made ship-to-ship radios soon available, and it is due to the presence of one on board the *Titanic* in 1912 that there were any survivors at all of that infamous tragedy.

As with the development of the battery and the electric generator, it is difficult to overstate the importance of the development of radio. Countless are the devices we have available to us today which use this kind of electricity, and many are the applications yet envisioned.

So Just What IS Electricity?

From the earliest days of Greek experimentation with static electricity to the much more recent experiments of Hertz and Marconi, physicists had struggled to understand what electricity was. The idea of a fluid was the only metaphor they had readily available, and although there were many things explained through this metaphor, there were many other things about electricity that were not explained by it. How could it flow in solid wire and in empty space? How could it propagate at the speed of light? Why did some materials conduct electricity well, while others either conducted it very poorly or apparently not at all? How could one tell how much electricity was stored in a Leyden jar? How could one tell how much electricity a battery would deliver? Just what WAS electricity?

The answers to these questions would have to await the discovery of the electron and the atom, which began in the early 1900s. We are again indebted to the Greeks for the idea of the atom. The word "atom" was used by the Greek philosopher Democritus to describe the smallest possible piece of matter that could be created by dividing a lump of matter into successively smaller pieces.

The history of the atom, as fascinating as it is, is not central to this discussion, and so will be gently side-stepped. Suffice it to say that some of the greatest minds of the 20th century worked on this, and their names (such as Einstein, Bohr, Heisenberg, and many, many more) became very familiar. The results of their work are immeasurably valuable to us, as they give us a way to answer the questions posed in the first paragraph of this section.

An atom, according to the Bohr model, consists of a nucleus (composed of protons and neutrons) and electrons which orbit the nucleus. As important as the nucleus is, we have no need to discuss it when trying to understand electricity. Indeed, even most of the electrons are of little importance to us. The only electrons important to us are the outermost electrons.

One of the ways that ordinary atoms of matter differ is in the number of electrons they prefer to have. For example, carbon prefers to have 6 electrons, oxygen prefers to have 8 electrons, silicon prefers 14, and copper prefers 29. But in our endeavor to understand electricity, even these numbers are not really important to us.

What matters to us is the *structure* of the electrons about the nucleus, and the behavior of one particular group of electrons known as the *valence shell*. Electrons can be thought of as orbiting the nucleus in discrete shells or concentric rings (see figure below); each shell can be thought of as a different distance from the nucleus. The outermost shell (the last one which has at least one electron in it) is known as the valence shell. Most atoms strongly prefer to have 8 electrons in their valence shell. If a particular atom has only one electron in its valence shell, it is relatively willing to give up that electron, and thus wind up with a full valence shell (the shell beneath the previous valence shell), which leaves the lonely single electron without a strong affinity for its original home.

The classic example of this type of atom is copper. The outermost two shells of copper have 8 and 1 electrons, respectively. Thus copper is very willing to give up its single electron in its valence shell to pretend to have a valence shell of 8 electrons. Since all the copper atoms feel this way about their single electron in the valence shell, there is a "sea of electrons" all running about without any particular affinity for one atom or the next. This is precisely what makes copper such a good conductor of electricity—this sea of electrons all very loosely bound to their atoms, ready to move around if given a bit of a push. This sea of electrons is common to the group of elements we know as metals, which group includes silver, copper, gold, aluminum, iron, chromium, nickel, zinc, and others. This is what makes these metals good conductors of electricity.

On the other end of the scale are atoms that normally have a full valence shell of 8 electrons. Our model would predict that, having 8 electrons, these atoms would seek to neither give up nor accept any electrons; there would be no sea of electrons. Thus we would predict

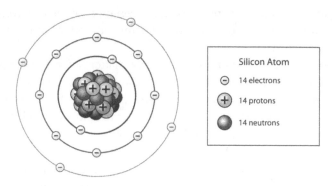

Figure 1-2:
Classical structure of the atom, showing a silicon atom with four valence electrons.

that these atoms would be very poor conductors of electricity, and this is indeed the case. These atoms are termed "inert", meaning that they do not react readily, and are good electrical insulators. Inert atoms include helium, neon, argon, and krypton.

In between these two ends of the scale are atoms that normally have 4 electrons in their valence shell, as in the silicon atom pictured in Figure 2. Four electrons means these atoms are neither good conductors nor good insulators; they could be called semi-insulators, but are more commonly known as semiconductors, and include the elements carbon, silicon, and germanium.

Now that we have a model of the atom that allows us to predict what elements will be good conductors, we still haven't quite explained what electricity *is*! A very useful metaphor for what electricity is would be a swarm of gnats in a breeze. The gnats represent electrons, all freely roaming about in the sea of electrons common to electrical conductors. If a breeze comes along, they continue to freely roam about but are driven as a group in a common direction. The breeze is the voltage, otherwise known as the electromotive force (i.e., the force which moves electrons). Two Volts pushed electricity exactly twice as hard as 1 Volt, just as a breeze of 2 mph would push on the swarm of gnats twice as hard as a breeze of 1 mph. But what is Voltage, or where do we get it?

Voltage is, as explained before, the force which causes electrons to move together in a common direction. The most common way of creating this voltage is by moving a magnet in the presence of a conductor (such as copper wire). The magnetism interacts with the negative charge of the electrons and makes them want to move. The number of electrons which participate in this flow of electrons determines the current (or amperage), which is measured in electrons/second just as the flow of water is measured in gallons/minute. So, if you were to watch the electrons in a wire go flying by and happened to notice that about 6.24×10^{18} of these electrons went by every second, you would say that 1 Ampere (or 1 Amp) of electricity was flowing (it takes a LOT of electrons to produce electricity, because they are so incredibly small!)

So there we have it; voltage causes the electrons to want to move along in the wire, and when a lot of them move, we call it current. The product of these two, voltage and current, is what gives us power, which we measure in Watts. For example, a 60-Watt light bulb is powered by 120 Volts and 0.5 Amps; the product of these two numbers gives us 60 Watts. A 100-Watt light bulb is powered by 120 Volts and 0.833 Amps.

What is it about electricity that produces light? In an incandescent light bulb, it is the heating of the filament. Many metals glow red when heated; if heated further, they will glow orange, yellow, and even white. The problem with metals that are white hot is that they become very brittle and break very easily; this is true of the tungsten filament in ordinary household light bulbs. If a lit lamp falls and does not break the glass of the bulb, it will still probably break the filament and burn out the bulb. If the lamp is not lit when it falls, the filament will probably not break.

But what is it about electricity flowing through a wire that produces heat? Heat is the movement of molecules; adding heat to something means causing its molecules to move more. It is admittedly strange to think of a solid piece of matter, such as the blade of a screwdriver, as having molecules that are moving around, since the screwdriver appears to be very solid to us. But even in the steel blade of a screwdriver, where the molecules are relatively tightly bound to each other, the molecules still rotate and vibrate over very small distances. Increase this movement sufficiently and the matter actually changes state; the extra movement causes the molecular bonds to weaken and the steel screwdriver blade melts! There are many ways to heat matter; one of these is to cause electrons to flow in the matter. Some of the energy of the flowing electrons is occasionally given up to neighboring atoms and molecules as they collide, much as hitting one pool ball with another causes some of the energy of one ball to be transferred to the other. So as a few electrons collide and slow down, they give their energy to the atoms and molecules of the matter in which they were flowing, causing the atoms and molecules to move more (that is, to heat up). If enough of these electrons collide, much heat is created. A light bulb has many trillions of electrons flowing through it, and they collide readily in the tungsten filament (much more than in the copper wiring). These many trillions of electrons provide the many billions of collisions necessary to heat up the wire, and it glows (or incandesces).

Static electricity is simply the buildup of lots of extra electrons on a surface or in a general area. Shuffling our feet as we walk across a carpet causes electrons to be moved from the carpet to our shoes, and the negative charge of these excess electrons moves to spread itself all around our body. This is because the electrons all have negative charge, and since like charges repel, they repel each other and try to move as far apart from each other as possible. Thus our finger (or any other part of our body) can act as a place for these charges to move to another place that has a more positive charge (or less negative). So if you charge yourself up by shuffling your feet, then touch someone else, they are probably much less charged than you, so the charge you are carrying suddenly discharges through a spark to that person. The term "static," as mentioned before, refers to the fact that these charges can build up for a long time, not going anywhere (remaining *static*), then suddenly discharging through a spark.

Where This History Leaves Us

The end of this history leaves us around the year 1900. At this time, we can generate and store static electricity, build useful batteries, generate and use electricity, and transmit wireless (radio) electromagnetic waves to send messages. By 1910, we know something about the electron, the atom, and the fundamental nature of matter. The only electrical devices that exist are simple electric motors, telegraph, telephone, the light bulb, batteries, and generators. Most cities and homes do not have electricity. While many advancements have clearly been made, the general public is without the benefit of any electrical devices.

It would be useful at this point to have a working definition of the difference between electricity and electronics. While there are arguably several, the one that will be the most useful for this discussion is this: *electricity* refers to devices that use the properties of electromagnetism, such as motors, relays, solenoids, generators, wire, antennas, and switches; *electronics* refers to devices that use the properties of the electron, such as vacuum tubes, transistors, and integrated circuits. Using this definition, this history leaves us knowing quite a bit about electricity, but almost nothing about electronics. This is exactly where we need to be to appreciate and understand the amazing developments that were soon to unfold.

Chapter Take-Aways

Our current model of the atom gives us a very workable understanding of what electricity *is*, certainly much better than was understood before that model was developed. So now after reading this chapter, each time you plug in or turn on something electrical, you will think of countless electrons, all being pushed by a moving magnet, flowing through the wire due to the voltage created by the moving magnet. Knowing this, it is easy to recognize that when the generator stops turning (moving the magnets), the electricity stops flowing—immediately.

This model also helps us understand part of what happens when a battery dies. A fully charged battery can deliver a certain number of electrons (a very LARGE number), but because electrical currents (even small currents) involve the flow of very large numbers of electrons, it doesn't take long for that supply of electrons to be depleted. Recharging a battery involves re-injecting electrons into the battery, which takes time (just as refilling your gas tank takes time).

And the next time you use your cell phone, you can think of the very small amounts of electricity that leave and enter your phone, carrying with them the data that your phone produces (mostly your voice or text messages) or uses (mostly other people's voice, or text messages, or data you download from the Internet).

All of these are examples of electricity in action, and without all these forms of electricity, and all the modern equipment we have to control its use, we would be back to the technology of 200 years ago—and that would be a huge change to our current lifestyles!

CHAPTER 2
THE VACUUM TUBE

Prior to the invention of the vacuum tube, the only way to switch electricity was either mechanically (as in an ordinary light switch) or electromechanically (as in a relay), and there was no way to amplify electric signals. This meant that there was no such thing as a public address system, no loudspeakers, no microphones, and no amplifiers. All speeches given prior to the invention of the vacuum tube had to be delivered at a volume that reached the entire audience; a speaker's voice could tire very quickly! Likewise, the only radios that existed could not be heard over a loudspeaker, but rather only with headphones (and even then they were difficult to hear!).

What Edison Learned About Electronics

Thomas Edison was the first to notice some of the electron effects that are unique to the vacuum tube. In the course of his work on the incandescent light bulb, he had learned much about how to build vacuum tubes (*a* vacuum tube is merely a glass tube which has been sealed after all the air has been removed; *the* vacuum tube generally refers to something that came later). Edison knew very well that any material that is heated sufficiently to incandesce will also oxidize (burn up!) if there is any air present, so all his work trying hundreds of materials for the filament of a light bulb was done with glass bulbs from which all the air had been removed (vacuum tubes).

After completing most of his work on the light bulb, Edison was struck by another idea: what would happen if he put another contact inside the light bulb? This question arose from his observation of what happened to his tungsten filament light bulbs. While the filament continued to burn for many hours, the bulb itself grew darker over time. Upon turning the bulb off, it was easy to see that the glass was darker on the inside. Analysis of the material causing the darkening showed that it was tungsten, which could only have come from the filament.

For making light bulbs, this darkening of the glass was clearly an undesirable occurrence, and much work was spent by many people over the years to reduce this problem. Yet even today, one can see on many well-used light bulbs that the glass has grown somewhat darker; we have not completely stopped the tungsten of the filament from coating the glass of the vacuum tube (light bulb). But how did the tungsten travel from the filament to the glass through the vacuum?

Figure 2-1:
Edison's vacuum tube diode, which was later developed by Fleming into a working product.

As Edison thought about this, he surmised that if he put another electrical contact inside the bulb, perhaps some kind of electric current might pass from the filament to the other contact. So, with the filament at the bottom of the bulb, he put another contact at the top of the bulb and watched to see what would happen. He was indeed able to measure a current, flowing from the filament to the other contact (later called the anode).

Prior to this, it had not been thought that electricity would flow in a vacuum. We knew electricity would flow in wires, and we knew we could create electromagnetic waves that would travel through the air or through a vacuum (radio waves). While Edison's observation was certainly a very significant finding, he did not come up with any useful applications of this phenomenon, nor did he succeed in explaining it. But Edison went one step further when he tried this same experiment with the polarity of the electricity reversed.

The results were quite unexpected; the electricity did *not* flow with the polarity reversed! Again, he did not come up with any useful application nor succeed in explaining this observation, but he did introduce it to the scientific world. This phenomenon of allowing electricity to flow in only one direction became known as the Edison effect.

Enter Fleming and DeForest

In the early 1900s, John Ambrose Fleming was working on detecting radio waves. The detectors then in use were very crude and had many other problems such as poor sensitivity and reliability, and an undesirable sensitivity to vibration, so he was searching for something better. Experiments with the Edison effect soon showed promise, and in 1904 he patented what he called an oscillation valve, which was also known as a thermionic valve, Fleming valve, or vacuum diode. It was substantially better at detecting radio waves, and soon became dominant in this application. It is also usually considered to be the defining device for the emergence of the field of electronics.

Shortly thereafter (1906), Lee DeForest improved on the Fleming valve by putting a small wire (later called a "grid") between the filament and the top contact (anode). This increased the sensitivity of the valve to high-frequency signals, and was later improved to function as an amplifier. The significance of this development was tremendous. For the first time in history, we could take a signal which was very small and amplify it to the point that it became useful in many ways. This vacuum tube, which DeForest called an "audion", later became known as a vacuum tube triode, and is what we can call *the* vacuum tube.

But how does a signal which is very small become larger? Doesn't this violate some kind of law of physics? If it doesn't, why can't we just make small things bigger? (How about that small lump of gold you

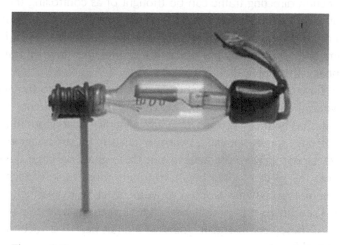

Figure 2-2:
DeForest's "audion", the first vacuum tube triode and the first amplifier.

keep on your dresser? How would you like to make it MUCH bigger?) If the signal itself were what became larger, it would be true that we had violated the known laws of physics; indeed, this is not what we actually do when we amplify a signal.

To amplify a signal, what we actually do is use the small signal to control a large, steady signal in such a way that the large steady signal becomes the same in shape as the small signal. This is best understood with an analogy.

Think of someone blowing air through a straw. We could take this straw and place over it a variable constriction which is driven by a lever and a small fan blade one could blow on. The fan blade is small, has very little mass, has precision bearings, and will begin to move in the slightest breeze. When this fan blade moves, it causes the variable constriction to close. With this setup, a person blowing very gently on the fan blade could easily constrict the straw, and thereby reduce the flow of air going through the straw. Thus a much larger flow of air (the air being blown through the straw) can be controlled by a very small flow of air (the person blowing on the fan blade). We have effectively amplified the blowing by causing the larger flow of air to be controlled by the smaller flow of air.

Another analogy is sometimes useful. A policeman standing in an intersection directing traffic can be thought of as controlling the flow of traffic with a force as small as that which it takes to move his hands. With this relatively small amount of force, he can stop the much larger automobiles and trucks. His ability to control the flow of traffic is essentially an amplification of force.

The flow of electricity from the filament (or cathode) to the anode, known as thermionic emission, is also a very interesting phenomenon. As discussed before, electrons in metals generally float about in a "sea of electrons", somewhat like a swarm of gnats in the air. Imagine what would happen if we were able to put these gnats on large doses of caffeine; the swarm would begin swarming more rapidly, each gnat moving in its somewhat random movements at an increasingly faster pace. Then imagine putting a large house vacuum near this swarm. Some of these very energetic gnats would get near this vacuum and get sucked into it.

18

The electrons in our cathode are given more energy by heating the cathode (more heat is adding more movement). The anode is put at a positive potential, which attracts the electrons (as the house vacuum attracts the gnats). In this setup, electrons "boil off" the cathode (thermionic emission) and are attracted over to the anode. The grid placed in between the cathode and the anode can be at either a positive or a negative potential, or no potential at all. And oh, by the way, potential means voltage. So, if the grid is at zero potential, it has no effect on this flow of current from cathode to anode by thermionic emission. But if the grid is at a positive potential, it *enhances* this flow, since the positive potential further attracts the electrons. And if the grid is at a negative potential, it *reduces* the flow from cathode to anode. So by applying a varying signal to the grid, we can cause the main flow from cathode to anode to be an exact copy of the smaller signal. As an example, the signal on the grid could be in the range of thousandths of a Volt ("milli"Volts), while the flow of current from cathode to anode could be 100 Volts. Using this much larger-amplitude copy of the smaller signal allows us to take the faint signal from a radio detector (in the milliVolt range or even smaller) and amplify it to the range that it can drive a loudspeaker (in the 200 Volt range), allowing all in the room to be able to hear the radio easily, without headphones.

Amplification also allows us to use a device which converts the human voice to a weak electric signal (a microphone), then amplify this weak signal, then send it hundreds of feet to loudspeakers placed around a large room or outdoor stadium (a public address or "PA" system). Vacuum tube amplifiers, together with microphones and loudspeakers, were the first products in history which allowed one person to easily address crowds of tens of thousands or more. What a huge improvement over shouting!

Future Uses

Vacuum tube amplifiers in radios and PA systems soon became widespread. Another application that soon became very popular was the phonograph. After that, the next major use was television, about 25 years later, followed by color television (see Chapter 6). And only about 10 years after that, vacuum tubes were used in the early days

Figure 2-3:
The Edison phonograph. Though very innovative, it lacked much volume, and its effectiveness and popularity were greatly enhanced by the vacuum tube amplifier.

of the computer (see Chapter 9). Suffice it to say that the vacuum tube amplifier was very successful in the early days of radio, television, and computers, in spite of its inherent limitations (more about that in Chapter 3).

Conclusion

The invention of the vacuum tube, which was improved upon for many years thereafter, finally gave us practical voice radios and PA systems. It was soon applied to the very weak sounds coming from gramophones (record players), allowing all to hear them easily, and replacing the familiar horn characteristic of early phonographs. It was also applied to commercial radio (AM radio), which began broadcasting in the 1920s, and to the phone system to amplify the small signals coming from each person's landline phone (see Chapter 11).

The vacuum tube was also applied with great success to the developing fields of electronic test equipment (meters for measuring voltage, current, wattage, etc.), radar, television, and later to the developing field of the computer. These will be covered in later chapters.

Until about 1950, the vacuum tube defined the field of electronics. After 1950 its popularity greatly declined in favor of the transistor (see next chapter), although the vacuum tube still remains popular in niche applications. The most

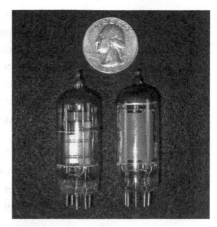

Figure 2-4:
Examples of vacuum tubes, compared in size to a US quarter (25¢ piece).

ubiquitous of these niche applications is the cathode-ray tube, or CRT, which until recently dominated as the viewed part of television and of computer monitors. Another modern vacuum tube is the very popular magnetron, the source of the microwaves that heats the food in essentially all microwave ovens. CRTs and magnetrons are highly modified and unique types of vacuum tubes.

The type of vacuum tube most often used in older tube amplifiers is still popular in high-end electric guitar amplifiers. Highly discriminating players of electric guitars will tell you, and emphatically so, that there is a substantial difference between the sound of a tube guitar amplifier and the sound of a transistor (AKA "solid state") guitar amplifier. This author has heard both types of amplifiers, and has studied the science between why they sound different; he is convinced these guitar players *know* what they are talking about!

Chapter Take-Aways

Amazingly, even though they're over 100 years old and have been replaced in most of their former applications, vacuum tubes are still with us in the very specialized applications mentioned above. And some of the most significant aspects about the vacuum tube are the many inventions that it enabled, all of which are still with us today, but in much-improved forms. Thus, today's telephones, amplifiers, televisions, computers, radios, and all other electronic devices all owe their existence to the humble vacuum tube.

In their heyday, vacuum tubes were manufactured by the millions. Because they wore out relatively often, it was not uncommon to find vacuum tube testers in the local drug stores, where customers could bring in their tubes and test them all, replacing the worn-out ones and thus fixing their television or radio themselves. Today, the market for vacuum tubes is a very small fraction of what it used to be, and that is unlikely to change. But we can all feel *very* grateful for the many things we enjoy today that were enabled by them!

CHAPTER 3
THE TRANSISTOR

The vacuum tube truly introduced the era of electronics, but it was not without its problems. Vacuum tubes only operated at relatively high temperatures (the cathode had to be heated to provide sufficient thermionic emission), which meant they were very hot to the touch while operating, and they produced lots of heat, all of which was essentially wasted. They did not last very long (most people from the vacuum tube era can remember going down to the local drug store to buy replacement tubes). They were fragile, since they were made of glass. And they could only be made so small before they were impractical (the smallest were about 1" high and about ¼" in diameter). They also took about 30 seconds before they warmed up enough to operate properly. Due to the amount of heat they produced and the energy needed to create this heat, they required substantial amounts of power. And finally, their upper frequency limit, which initially seemed plenty good, soon became a limitation as radio and radar pushed into continually higher frequencies. It was research into how to solve this last problem that eventually led to the development of the transistor.

Work at Bell Laboratories
While there were several institutions which were working on developments in radio and radar (which originally was an acronym standing for RAdio Detection And Ranging), it was the work at Bell Laboratories that is usually most recognized for the invention of the transistor. Some kind of detector was needed that would work at the high frequencies required by radar; the vacuum tube diode was not able to detect these very high frequencies. Even the vacuum triode, and subsequent improvements such as the vacuum tube tetrode, pentode, and other variations, were not meeting the needs of radar. On a moment's inspiration, one worker decided to try one of the original radio detectors from decades before, a "cat's whisker" detector.

These were made of a crystal contacted by a tungsten wire (the "cat's whisker"), and during the era of their widespread use in crystal radios (prior to the vacuum tube diode and triode) their operation was not understood at all.

Later research would reveal that a crystal radio detector was simply an early type of "solid-state" (as opposed to gaseous or vacuum-state) diode. We remember from the previous chapter that a diode was very useful as a radio detector, and was also the precursor to the triode, the first amplifier. At Bell Labs, this crystal detector turned out to be fairly successful at detecting radar, so a great deal of research began into finding out how these detectors worked.

One of the characteristics of crystal detectors was that they were very finicky. Early radio listeners put on their headphones and began carefully moving the cat's whisker over the surface of the crystal. Eventually they would find a place where the signal would come in well, and listening could be enjoyed without further intervention. Unfortunately, this "magic" spot did not last, and in a few hours a new spot would need to be found. Also, sometimes it would be very difficult to find a magic spot at all. Though they were temperamental, they still worked, and were used widely in early radio because they were the best option available.

This tendency to be temperamental was clearly a problem that had to be solved if this crystal detector was to be used in radar. The purity of the crystal, and of the tungsten, were early areas of research. The purity of the crystal was found to be a much more significant factor, so a great deal of research went into how to grow and purify crystals. This engrossed many research organizations and took many years, but steady progress was made. And as the crystals became purer, the successes became more frequent; the momentum was beginning to build.

It is also interesting to read about some of the things that were tried to learn about these solid-state diodes. Because they were so finicky, and because no one had a theory how they worked, all sorts of ideas were tried. The temperature was raised and lowered. They were tried in dry air, in humid air, and even underwater. They were tried at high pressure and at low pressure (in a vacuum). They were held rock steady and they were vibrated. They were even tested in the dark and in bright light.

Surprisingly, this last particular test showed some very interesting results. A working crystal detector was subjected to bright light; in these conditions, the conductance of the crystal increased (it conducted electricity better). And it was repeatable; as the amount of light was lowered, the conductance went back down. Observing this phenomenon, Walter Brattain realized something that became central to their understanding of how a solid-state diode operated. To understand what he realized, we will need to learn something about crystals, especially very pure crystals.

The Nature of Crystals

History seems to have endowed crystals with a legend of magical powers. It is not these powers that are of any importance to us in this discussion, however. Crystals are one of three main molecular organizations that solid matter can be in. These organizations are commonly referred to as crystalline, polycrystalline, and amorphous. A diagram of these is shown in the figure below.

Amorphous materials are characterized by relatively random organization of the molecules; examples include glass, most plastics, and many types of ceramics and porcelain. Crystalline materials are characterized by an extremely consistent and regular pattern of organization of the molecules; examples include nearly all precious stones, such as diamonds, rubies, emeralds, and sapphires. Polycrystalline materials have many areas (called "grains") where there is a crystalline organization of the molecules, but these areas are relatively small, and there is

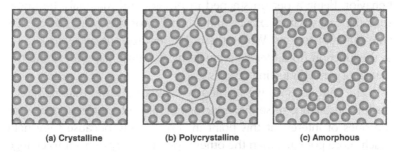

(a) Crystalline   (b) Polycrystalline   (c) Amorphous

Figure 3-1:
The three main organizations of solid matter.

significant irregularity at the interface between these areas of crystallinity. Examples include most metals, particularly iron and steel.

Diamonds are probably the most famous of crystals, and are certainly the most valuable. They are made of carbon atoms, bound together in a very tight organization, such that diamond crystals are the hardest material known to man. The quality of a gem diamond is primarily dependent on its size, clarity, inclusions, and color. The clarity of a diamond is dependent primarily on its purity and the consistency of its crystalline structure. Inclusions are small bits of foreign matter, included in the diamond as a result of the process of its formation inside the earth under intense heat and pressure. Inclusions are not desirable in gem diamonds. The color of a diamond is a function of the presence of minute amounts of other elements in the crystal, far too small to affect the clarity or to be considered inclusions. If one could make or find a perfectly pure diamond, it would consist only of carbon atoms, all equally spaced; it would be perfectly clear and transparent.

In time, crystalline materials would inherently be of more interest in the realm of electronics because their electrical behavior is much more consistent, predictable, and reliable. If a given crystal is not behaving in this manner, it is almost certainly due to the presence of other materials in the crystal (impurities), or due to irregularities in the crystalline structure.

The Magic at the Junction

What Walter Brattain realized was that there must have been some sort of junction operating in the crystal that exhibited the light-sensitive behavior. Being a person steeped in physics and materials, he knew something of what to expect, and this behavior was typical of crystals with different regions. Where these different regions met was known as the junction. To understand junctions, we must first understand the nature of semiconducting crystals.

In the Introduction, we learned what makes some materials conductors, insulators, or semiconductors. Semiconductor materials were the ones of interest at this point in time, simply because they held much more potential than the other types. It is very difficult to change the electrical behavior of a conductor (to make it act more like an insulator) or of an insulator (to make it act more like a conductor).

However, it is not very difficult to change the electrical behavior of a semiconductor, since it lies somewhere between the two extremes of conductors and insulators.

The property of resistivity is of great importance in working with semiconductors. This somewhat unusual term is actually quite easy to understand. It refers to the inherent resistance offered by a piece of a given material when that piece is a cube (for example, 1 cm on each side). There is no other property of materials which ranges so widely as resistivity. For example, the resistivity of a 1-cm cube of polytetraflouroethylene (also known as PTFE or Teflon®) is in the range of 10^{18} Ohms, while the resistivity of a 1-cm cube of silver is in the range of 10^{-5} Ohms. These materials are essentially at the extreme ends of this property of resistivity; notice that it ranges by about 23 orders of magnitude (10^{23})!

If we were to separate materials into their three main families, according to their resistivity, we would have conductors at the low end, with resistivities ranging from 10^{-5} to about 10^2. Next would come semiconductors, with resistivities ranging from 10^3 to about 10^9. The last group would be insulators, with resistivities ranging from 10^{10} to about 10^{18}. These divisions are approximate, but give an understanding of where semiconductors lie.

If we take a 1-cm cube of crystalline silicon, all the atoms will have 4 electrons in their valence shell (see Figure 1-2), and it will have a resistivity of about 2.4×10^5 Ohms. This resistivity can be greatly altered by a process called doping, in which a few silicon atoms are replaced by atoms with only 3 electrons in their valence shell (leaving too few electrons; known as *p-doping*), or by atoms with 5 electrons in their valence shell (having extra electrons; known as *n-doping*).

Having taken the long way around, we finally come back to the topic of this section: the magic at the junction. If we create a region in this silicon crystal in which some of this p-doped material is directly in contact with some of this n-doped material, something very important happens. The extra electrons in the n-doped material fill all the holes in the p-doped material, giving the junction region a very definite preference for which direction current will flow. If we attempt to continue this movement of electrons from the n-doped area to the p-doped area (by applying a negative voltage to the n-doped area or a positive voltage to the p-doped area), this movement will proceed with little opposition. If we attempt to reverse this movement of electrons, it will be opposed greatly. We have created a diode, operating on solid-state physics instead of vacuum physics, but still permitting only one direction of current just as the vacuum-tube diode did.

The foregoing was explained based on much of what we know today about solid-state diodes. When Brattain realized there was some kind of junction at play in this unusual crystal, we knew very little about the behavior of junctions. Quickly, a great deal of effort was poured into learning about the nature of these junctions and how they could be produced and made consistent. The cat-whisker crystal detector was later dubbed the point-contact junction, and later the Schottky junction. It was notoriously finicky; after years of lab work, even the best ones would not last long, and only a small percentage of the ones that were produced actually worked.

Doubling the Magic

Following on the example of the triode, the physicists and engineers knew that if they could put another junction in the material, and use it to *control* the unidirectional flow, they would have a solid-state amplifier and switch, capable of replacing the vacuum tube. Although they probably did not fully comprehend the vast impact such a device would have in the coming decades, they did realize that it would be a major advance in the field of electronics, so they worked hard and long. The pioneers that generally receive historical credit (there were certainly many others) are Walter Brattain, John Bardeen, and William Shockley, all of Bell Labs. They received the 1956 Nobel Prize in Physics for their invention of the transistor.

After making several devices that briefly appeared to exhibit amplification, they finally were able to produce one that remained consistent long enough to be shown to their superiors. Their lab books record the date as December 23, 1947. It was a point-contact type of transistor (later known as a field-effect transistor), and still suffered from problems with inconsistent behavior.

The next years were devoted to making purer crystals of germanium (the semiconductor material they were using at the time) and to solving the problems inherent in trying to produce these transistors in a reasonable volume and at a reasonable price. These problems were not insignificant, and due to the rather poor quality of the early transistors, their very limited volume and their high price, they did not catch on very quickly.

As would be expected, Bell Labs was not the only company working on this very important development. At Texas Instruments, the problems with germanium were just as persistent, so they decided

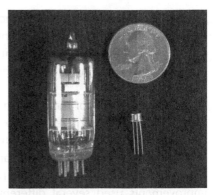

Figure 3-2:
A vacuum tube alongside a transistor, demonstrating the dramatic size difference.

to try using silicon. It turned out to be an excellent choice, as silicon was easier to purify and to process in the other steps required to make a transistor. By 1954, Texas Instruments was able to produce transistors in sufficient volume and quality to attract the attention of electronic designers throughout the industry. Although at first not cheaper than vacuum tubes, they soon became so; they were also much smaller, used much less power, were more rugged, required no warm-up time, had a higher frequency limit, weighed less, and used lower voltages. In nearly every way they were superior to vacuum tubes, and in the ways they were inferior, designers soon found ways to work around the inferiorities. By about 1960, most new products were being designed to use transistors; only a few years later, only niche applications remained for vacuum tubes. The transistor appeared set to make a clean sweep and relegate the vacuum tube to being only a fond memory.

More Magic at the Junction: Solar Cells

The very interesting light-sensitivity of the crystal noted by workers at Bell Labs did not remain simply a lab curiosity. Work as early as 1839 had shown some very interesting but inexplicable results using various metals and solutions—an electric current was produced when they were exposed to light. Later work (1877) used the semiconductor selenium with the metal gold; work in 1930 used the semiconductor copper oxide with copper. All of these produced an electric current when exposed to light, but were so inefficient that they were not practical as significant sources of electricity. However, they were found to be useful for measuring light, and the field of light meters moved to solid state.

The understanding of the atom that developed in the early 1900s finally gave us the required theory of operation, and much about these junctions was learned. Eventually, improvements led to the development of a silicon solar cell, around 1940. This was significant because silicon was inexpensive and relatively easy to work with. Later improvements used several different types of semiconductors, and raised their efficiency. By 1960, the solar cell was a practical alternative for a source of electricity in remote locations. For example, most of us have seen pictures of satellites with their solar panel "wings", supplying the power for the electronics of the satellite.

Today, solar panels have developed to the point that they are beginning to power buildings (especially the highly "green" buildings receiving LEED certification (Leadership in Energy and Environmental Design). They are also used in many photovoltaic (PV) power plants, the largest of which generate nearly 100 MW. In the relatively near future, they appear to be able to become affordable enough to serve as roofing on homes, thus reducing or eliminating homeowners' power bills. Such PV power generation is looked to by many as one of the best ways to free the world from our dependence on ever-dwindling supplies of fossil fuels.

Conclusion

Time would come to show that, in most ways, the vacuum was indeed relegated to only a memory. The transistor soon dominated for applications in communications, consumer products, test equipment, military products, and eventually computing. "Transistor" even became the moniker of one very significant product it enabled: the portable transistor radio. It was the first truly portable consumer product, and became very popular.

The idea of portability was not new, yet until the transistor it was not very practical. The transistor enabled portability due to the much reduced size, weight, and power consumption of the transistor, in

Figure 3-3:
Battery sizes D through AAAA, N, and a button cell battery. Note the missing A-size battery, which has not been sold for many years.

addition to its greater ruggedness. The transistor radio became a 1960s status symbol, of having the latest thing. It also greatly improved the car radio, to the delight of many car owners.

To be truly portable also meant that it had to be battery-powered, which sparked innovation in batteries. Previously, small batteries had come in basically one size (D), one type (carbon), and one voltage (1.5 Volts), and were used almost exclusively to power flashlights. Soon they came in several smaller sizes (C, B, A, AA, AAA, AAAA, N, button cells, etc.), several types (alkaline, nickel-cadmium, others), and several voltages (9 Volts, 1.2 Volts, others). Together with transistors, this soon led to the development of other portable devices, including hearing aids, phonographs, cassette tape players, televisions, and remote controls for television. Satellites also became practical, especially when paired with solar cells. In many ways, it was the dawn of smaller, less expensive, and more practical electronic products.

The next chapter will discuss the integrated circuit, which allowed for many transistors to be put together on a single substrate, and began the process of integration which has continued unabated for the past five decades. But there is still a rather large market for discrete (individual) transistors in special applications, and it does not appear that this will change.

Chapter Take-Aways

The transistor enabled portability; without it, we would not have our portable devices such as cell phones, laptop computers, tablet computers, GPS navigation guides, and a host of other devices that we often ignore. And the transistor is more popular than ever. Last year, over 100 billion of them were made for every person on the planet—that's over 700×10^{18} transistors—in just one year! A typical cell phone today has over 10 billion; a typical tablet computer has over 50 billion; a typical laptop computer has over 100 billion. Cars have billions; televisions, radios, CD/DVD/Blu-ray players all have billions; video games have tens of billions; even a sewing machine has tens of millions of transistors. No item in the history of the world has ever been manufactured in such large volumes as the transistor is today, and this is a trend that is likely to continue.

Where once the transistor was new and the public star of the show, this is only true today for those who design the circuits that use them. Transistors are still the star of the show in electronics, but today's general public has no idea they are carrying so *many* of these with them every day. It is, without a doubt, the most ubiquitous item created in the 20th century, and possibly in the history of the world.

CHAPTER 4
THE INTEGRATED CIRCUIT

The transistor (Chapter 3) was truly a major advancement in the field of electronics, and enabled many rapid advances. However, had the magic of electronics ended with the invention and eventual success of the transistor, most of what we have today in electronics would still not exist. If it were not for the invention of the integrated circuit and the subsequent unremitting march of what came to be known as "Moore's law", we would have very little beyond the products described in the previous chapter. And while those products were significant advances over what we had before, they pale in comparison to the products and advances that have marked the past several decades since the invention of the integrated circuit.

In the annals of the history of technology, it is not unusual to find a new idea emerging in more than one place and in more than one form at the same period of time. When historians have noted this, they have often termed these as ideas "whose time had come". Such is undoubtedly the case in the invention of the integrated circuit.

Only What Was Essential

Modern electronic circuits include many different elements, including many devices which are generally unfamiliar to most people. These devices include relays, solenoids, motors, transducers, igniters, switches, optical detectors, optical emitters (lights), varistors, potentiometers, rheostats, connectors, crystals, cables, sockets, inductors, fuses, transformers, and still other devices. But most of these elements are special-purpose devices, and are used only in relatively limited numbers. The main work of electronics, which involves amplification and switching, can be done using only three devices: transistors, resistors, and capacitors.

Designers of electronic circuits have long known that most electronic circuits can be designed with the right combination of these three devices. Transistors are the star of the show, the workhorse of

33

electronic circuits; more transistors are made each year than any other electronic device. Resistors and capacitors are the setup crew and the warm-up act for the star of the show; essential, but not getting much of the glory. With the invention of the integrated circuit, this trio would become essentially the only members of the broad family of electronic devices to be included. Other devices would eventually be worked on for inclusion in integrated circuits, but since the majority of electronic functions could be designed with just these three devices, it was enough that they were included in integrated circuits.

Transistors have been addressed in the previous chapter. The function of these other two elements, resistors and capacitors, is a little more difficult to appreciate without more background, which is not the purpose of this book. Suffice it to say that without resistors and capacitors, there are many things which transistors would be very poor at doing, or could not do at all. But with these mostly humble devices, together with the transistor, the world of possibilities truly opens up.

The essential part of a resistor is just what its name implies: resistance. Not just any amount, but an exact amount of resistance, allowing designers to determine what currents and voltages will exist, and where. Typical resistors have values from about 10 Ohms to 30 Mega (million) Ohms. In Chapter 3, we reviewed the nature of semiconductive materials and pointed out that they have a resistivity somewhere between 10^3 and 10^9 Ohms. This means that semiconductive materials would work quite well for making common resistors.

Capacitors are very simple devices which store a small amount of electrostatic charge, as the Leyden jar did. The essential parts of a capacitor are two conductive plates or surfaces, separated by a thin insulator. This is how your body is able to store a charge, just as a capacitor does: your body (which conducts electricity fairly well) is one of these conductive surfaces; the earth (or the floor) is the other. The sole of your shoe is the thin insulator which separates the two surfaces. Just for fun, on a nice dry day when you can easily generate static electricity, note how removing this thin insulator (going barefoot) eliminates this capacitor, and you can no longer generate static electricity!

Electronic circuits have always had these two parts of a capacitor: conductive materials and insulative materials. Copper is the most common conductive material; insulative materials include the insulation around the wire, and the air itself. To make an IC, we had to have materials to connect to the transistors (wire) and insulative materials to keep these connections from shorting to places they should not. Thus, in electronic circuits, all the elements were there: transistors, resistors, and capacitors, and conductors and insulators. But before the integrated circuit, the problem was that these elements were each separate, discrete elements in the circuit, so the circuits could only be made so small before it became unmanageable to build the circuit.

Two Places At One Time

For the invention of the integrated circuit, history gives credit to two individuals and two companies, both working independently and with different approaches. Jack Kilby had recently been hired at Texas Instruments; Robert Noyce was a co-founder of Fairchild Semiconductor. Both came up with remarkably similar solutions to the problem of making electronic circuits smaller. Kilby and Texas Instruments filed on Feb 6, 1959, and were awarded U.S. patent #3,138,743 on June 23, 1964. Noyce and Fairchild Semiconductor filed on July 30, 1959, and were awarded U.S. patent #2,9181,877 on April 25, 1961. Through cross-licensing agreements, the industry grew steadily to over $1 trillion today, a phenomenal growth by any measure.

The number of things an electronic circuit can do is directly a function of the number of devices it can have, particularly the transistors (the star of the show). The first integrated circuit had one transistor, three resistors, and one capacitor, and measured about 1" by ½". Both companies began designing and building circuits with more than just basic, simple functions; and both companies found numerable problems that needed to be solved. By 1961, commercial integrated circuits became available. These circuits had up to 10 transistors, along with associated resistors and capacitors where necessary. This stage was known as SSI, or small-scale integration.

Early integrated circuits (ICs) were very expensive to produce, but the military and our accelerated space program had great need for these much smaller products. The military was in the middle of the Cold War, and the missiles being planned needed complicated yet small guidance systems, which could only be produced using ICs. NASA was going full blast toward the goal of putting a man on the moon by the end of the

1960s (which they succeeded in doing in July of 1969); such an ambitious goal could not have been reached without ICs.

Soon the complexity of ICs began to be measured only by the number of transistors they contained. Designers learned ways to avoid using capacitors when possible (they took up a lot of valuable room, or "real estate," on the integrated circuit), and resistors were used only when transistors would not suffice. By 1965, ICs were available with up to 100 transistors, a stage known as medium-scale integration (MSI). Only a few years later (1970) they had moved to large-scale integration (LSI), using up to 1,000 transistors.

The naming of eras or stages in IC complexity reached VLSI (very large-scale integration) in about 1975, with up to 10,000 transistors. Some coined another term for the era with up to 100,000 transistors (ULSI, for ultra-large-scale integration), but the term VLSI, for most people in the industry, seemed to stick the best. Indeed, the fascinating field of IC design is often termed VLSI even today, although we now have ICs with over 100 billion transistors.

Moore's Law

One of the co-founders of Intel, Gordon Moore, observed in 1965 that the number of transistors per square inch of silicon had doubled every year since the invention of the IC, and he predicted that this trend would continue. Although the pace eventually slowed to a doubling every 18 months, and lately to a doubling every 2 years, this inexorable march of progress in ICs has been dubbed "Moore's law". Although there is no law to it at all (it should have been called Moore's prediction or observation), it has held true for over 4 decades, with the aforementioned reductions in the rate of doubling. This continual progress has made possible something this generation has grown to take for granted, but which would be impossible without the IC and the steady march of "Moore's Law".

As mentioned in Chapter 3, the transistor had several advantages over the vacuum tube. The transistor:

- was smaller
- used less power
- was more rugged

- operated at higher frequencies
- was lighter (weighed less)
- eventually was less expensive
- eventually was more reliable.

Any one of these advantages would have been enough to have guaranteed the transistor a rather permanent and important existence. Two or three of these advantages would soon have given it the major portion of the growing field of electronics. But the fact that it eventually had seven major advantages over the vacuum tube guaranteed the demise of the vacuum tube and the rise of the transistor.

The IC did this all over again, in comparison to the transistor. It was smaller; it used less power. It was more rugged (than the equivalent circuit made of discrete transistors) and it could operate at even higher frequencies. It weighed less, eventually became less expensive, and eventually was more reliable than the equivalent circuit made of discrete transistors. Clearly all the magic that the transistor had wrought on designers and the willing public was going to occur again, and this time it appeared that the only thing necessary to continue the miracle was a continuation of Moore's law. With every new generation of IC, the new circuits achieved all seven advantages over the previous decade of ICs, because of this continued progress known as Moore's law.

The progress in ICs has often been compared to progress in automobiles, primarily to illustrate the dramatic nature of what has happened. This author greatly enjoys this comparison, so it bears repeating. In this comparison, it is acknowledged that some of

Figure 4-1:
An integrated circuit (IC) alongside a transistor and a vacuum tube. The IC pictured has over 256,000 transistors, yet is considered VERY old by today's standards.

the comparisons are somewhat meaningless, as some of the limits on automobiles are set by the size of the occupants, which is not subject to being successively shrunk. But if we gleefully ignore these practical constraints, we can readily say that, if we had improved the automobile of 1960 as much as we have improved the IC of 1960, today's car would:

- have dimensions of about 600 microinches by 1000 microinches, which is much smaller than a grain of salt;
- use so little gasoline that 1 gallon of gas could easily last more than 1 billion miles;
- be so rugged that it could easily withstand being dropped from the top of the Empire State Building, run over by an 18-wheeler, and driven across the Baja Desert at 300 mph;
- run at up to 300,000 rpm;
- weigh less than ¼ ounce;
- cost less than 10 cents;
- last for more than 10 million miles.

While these comparisons may seem preposterous, this author has done the calculations himself, and they are accurate. Wouldn't it be great to pay 30 cents (for a very nice, luxury car), put in a gallon of gas and drive it all your life (and your children's lives, your grandchildren's lives, your great-grandchildren's and so on for 50,000 years, assuming 20,000 miles per year), on the same gallon of gas! Of course, on the impractical side, weighing only ¼ ounce, it would blow away in the slightest breeze and be impossible to find in the parking lot (or anywhere else, for that matter!)

But the point of this somewhat inane exercise is not to disparage advances in automotive transportation (which have been dramatic), nor to move to the ridiculous, but to allow us some comprehensible form of comparison. The decades of improvement on the IC have caused us to expect next year's electronic products (such as cell phones or computers) to be smaller, lighter, use less power, work faster, do more, and cost less. This type of expectation is *only* true in electronic products, and this only due to the steady onward march of Moore's law.

The Magic of Photolithography

The process of making ICs is one of the most demanding, complicated, time consuming, expensive, and super-high-precision processes ever known. Let's consider the case of the average IC. The average IC costs about $8 million to design and build; roughly $6.5 million of this are the design costs alone, since such design work is only done by highly trained electronic circuit designers. The average IC brings in an average of $32 million, an excellent return on investment (if it sells!) It is made of 20 layers of carefully added and selectively removed extremely pure materials, in about 50 different operations, some of which are repeated several times; the total number of steps or separate operations for making it is about 500. It takes about 5 months to manufacture it.

The typical IC facility costs over $1 billion to build, another $1 billion to fill with incredibly complex and expensive manufacturing equipment, and another $150 million/year for the payroll of the highly trained people who work there. Yet despite all these enormous expenses, designing and building ICs is generally very profitable, and is at the heart of the $1 trillion/year electronics industry, the largest industry sector in the world.

At the heart of the process of making ICs is *photolithography*. The name itself contains the clues as to the meaning of the word: *photo* refers to light, as in *photograph*; *litho* is from the Greek word for stone: *lithos; graph* refers to making images or lines. Together, these terms mean using light to make images or lines on stone.

The most frequently used semiconductor material in the world is silicon. It is also the second most common material in the world, second only to oxygen. This is because the vast majority of this earth is made up of rock, sand, and dirt, the major component of which is SiO_2, also known as silicon oxide. Two parts oxygen for each part silicon—and there's a LOT of it! (The mass of the Earth is about 5.97×10^{24} kilos, or 6.57×10^{21} tons!) So it is not at all a stretch to say that raw silicon is as cheap as dirt, since that's where it comes from! It is also a good way to remember the meaning of the *litho* part of photolithography—silicon is the stone on which we make images or lines, using light.

The basic process of photolithography is quite old; the rudiments of it were known as far back as the early 1800s. There are many liquids and solids which change when exposed to light; some change color, some harden, others soften. And the ones most important in IC manufacturing are the ones that change permanently. These materials are known as *photoresists,* and because they either harden or soften when exposed to light, they are very useful in photolithography.

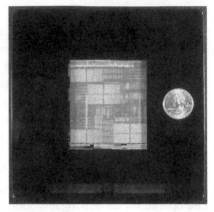

Figure 4-2:
A mask (known in the industry usually as a "reticle") used in the IC manufacturing process.

One way to describe the process of photolithography is to use a process with which most of us are familiar: painting a wall in our house. However, for this comparison to be useful, we will need to be somewhat *avant-garde* in our approach; we don't want to just paint the wall, we want to create very unique patterns, in a way no house painter has ever done. We want to apply the paint only in certain areas; we also want to remove part of the wall itself (only a thin layer) and put dyes into or pigments over these sections where we have removed part of the wall. We will end up with some of our final pattern being buried under layers of subsequent patterns, yet still being visible because the subsequent layers are extremely thin.

To accomplish this very unique method of painting, we first bring in our *photopolymer,* a form of liquid plastic which we paint over the entire wall. It dries when exposed to air, but does not harden, remaining soft like a thin layer of rubber. We will also need a special tool called a mask, much like a costume mask or a stencil: it has openings in the right places and of the right shapes, as defined by the pattern we want on the wall. After the photopolymer dries, we then fix the mask and a bright light so that they will not move, and shine the bright light through the mask onto the photopolymer. The photopolymer then hardens where the light shines on it. Then we scrub the wall with a gentle chemical which dissolves the *un*hardened photopolymer. We are then left with portions of the wall being covered by a thin, hardened layer of photopolymer (where the light shone), while other portions of the wall (where the light was blocked by the mask) have no photopolymer on them. We can then spray paint with the first color we want; the paint we spray on (blue, for example) will cover the entire wall (see Figure 4-3).

Now we have a blue wall. By next applying a chemical that can penetrate the paint without dissolving it, we attack the interface between the hardened photopolymer and the wall, causing the hardened photopolymer to fall off the wall, with its paint still on. What we are left with on the wall are some very sharply defined areas of blue paint. These areas were defined by shining the bright light through the mask onto the photopolymer. This is basically the process of photolithography, but we did it on a wall instead of on silicon (we would probably call this form of avant-garde art *photowallography*.)

This explains how we add paint to selected areas of the wall, but does not explain how we selectively remove material from the wall. Selectively removing material is accomplished by again using our photowallography process to first define the areas where we wish to remove material. Then we simply apply a chemical to the wall that will etch away the material we wish to remove, but will not etch the hardened photopolymer. After finishing this wall etching process, we remove the photopolymer as before.

These two processes using photopolymers allow for selectively adding or removing in defined areas of the wall. By repeating these processes as desired, we can build up layers on the wall, some layers actually lying beneath the surface of the wall, others lying on top of the wall, or even on top of other layers.

In making ICs, we use photolithography to define the areas where we wish to add or remove material. Processes used to remove materials include wet etching and dry (or plasma) etching; processes used to add materials include doping, ion implantation, chemical vapor deposition, sputtering, and evaporation. Which process is used depends on the material to be removed or added. When we are done, we have created several layers of areas which have been selectively defined by

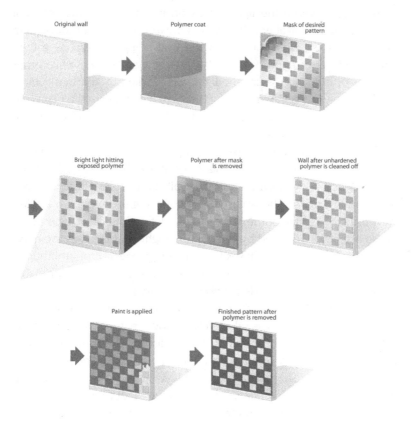

Figure 4-3:
An illustration of a wall being painted and selectively patterned using the "photo-wallography" process described.

photolithography, and which together make up transistors, resistors, capacitors, and the wires necessary to interconnect them. Below is an amazing picture taken by IBM of a very small area of one of their ICs, showing the layers of materials brought together to make a pair of transistors. Each area is carefully defined through photolithography, then the necessary materials are added as described above in our description of photowallography. This picture also shows some materials which were added to a layer beneath the surface of the original

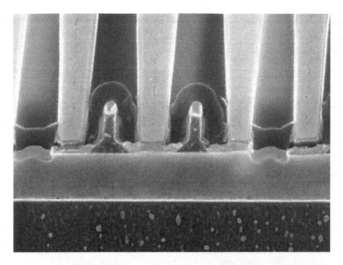

Figure 4-4:
A pair of complementary metal-oxide semiconductor (CMOS) transistors, also showing the first layer of wiring. To the left and right of this pair of transistors are portions of other pairs of identical transistors.

silicon, through a process of selectively removing silicon, then adding the desired material where the silicon was removed.[1]

How Low Can You Go?

One of the wonderful things about the process of photolithography is that we can make things so incredibly small. In the example we gave of photowallography, it is easy to see that one limit to how small we can make the areas on our wall is related to how small we can make the openings in our mask. This limit was basically removed many years ago, as we learned how to make masks with features as small as 0.5 nm,

---

[1] For those familiar enough to recognize this photo as the SOI process: I admit that this description is oversimplified (ignores the insulating layer on top of the silicon) and even a tiny bit erroneous (refers to the insulating layer as silicon), but for simplicity's sake, and the gist of the IC manufacturing process, this explanation should suffice.

which is about the distance between two silicon atoms—much smaller than we could make the IC features. Basically, we can make masks of much higher resolution than we can practically use. So if the limit isn't how fine we can make the openings in the mask, what is the limit?

The ultimate limit, for now, is the wavelength of the light we use to expose the photopolymer. A fundamental law of physics says that we cannot make the resolution (distance between the lines) any better (smaller) than one-half the wavelength of the light used. Visible light has wavelengths from about 700 nm (red light), up through orange, yellow, green, blue, indigo, and violet light, with a wavelength of about 400 nm.[2] If we use visible light with a wavelength of 400 nm, then in theory we can make features (areas) as small as 200 nm. Practical limitations actually stop us short of the fundamental limit, and restrict us to features about two-thirds of the wavelength of light, or about 267 nm using the shortest wavelength of visible light.

Of course, violet light is by no means the light with the shortest wavelength; there are wavelengths shorter than this at the higher frequencies above violet (ultraviolet); these are the frequencies of light that are the most damaging to the eye or cause sunburn. But there are some very serious challenges to using these frequencies, and practical limitations have restricted us to using ultraviolet wavelengths of about 150 nm, resulting in features as small as 100 nm. A great deal of time and money has gone into improving this (making things even smaller using even shorter wavelengths of light), but the challenges loom ever bigger with each generation.

The very early generations of IC photolithography used features around 10,000 nm, (also given often as 10 µm, 10 microns, or 10 micrometers). As Moore's law marched on, successive generations about 18 months apart each reduced this by about 1.5 times, so that the next generation was about 7 µm, followed by 5 µm, etc. Today's high-volume ICs are produced using features between 0.13 µm and 0.25 µm, depending on the need. This gives us anywhere from tens of thousands of transistors to several billion transistors on a single IC,

---

[2] A nm, or nanometer, is one-billionth of a meter, and is the most common unit of measure for wavelengths of light. 400 nm is about one-one hundredth the thickness of human hair.

depending on its resolution and size. And using workarounds and even shorter wavelengths of light, we have dramatically extended our reach, until now today's smallest features (in development) come in at about 25 nm!

How long can we continue to shrink the transistors, and thus keep Moore's law working? Present predictions show a path out to about 10 years from now, maybe 15. While this seems alarming, comfort can be found in the fact that for the past 3 decades, we have always been about 15 years away from ending Moore's law. Each time we have approached what seems an impenetrable barrier, some technology has been developed which has allowed us to work around or solve the problem. There is certainly a great deal of work and money presently being spent in finding ways to overcome the barriers we face today. The IC industry spends tens of billions of dollars and hundreds of thousands of person-hours each year on solving these problems. Will Moore's law eventually become impossible to continue? Yes, inevitably. But the magic continues today, and apparently will continue for at least several years to come.

What About Those Clean Rooms?

Occasionally, one will see pictures or short video clips of the inside of an IC factory, especially the areas where the actual production is done. These areas are known as clean rooms, and "clean" is truly the understatement of the year! They are cleaner than any other environment on the earth. They are over 1,000 times cleaner than the cleanest operating room in a modern hospital. They are thousands of times cleaner than the cleanest air atop a remote mountain in an untouched country. But they are not cheap!

Today's typical unfurnished 1-family house costs between $80 and $120 per square foot, depending on features. By contrast, typical cleanrooms cost about $10,000 per square foot! Why do they have to be so clean? Because tiny specs of dust, pollen, smoke, or other airborne particles are many times bigger than the features on an IC! Back on our example of photowallography, imagine how big the problem would be if the air were full of particles as big as the room we're painting! They would make our work impossibly difficult. And so in IC clean rooms, the air is super-filtered to remove particles down to about 0.1 µm, and they must do so very effectively. Clean rooms are rated by the number of particles per cubic foot that remain in the air; a class 1000 clean room has no more than 1,000 particles/ft³. The lower the number, the cleaner the clean room (and the more expensive to build!) Modern clean rooms are usually class 10, and often have microenvironments within them of even lower (cleaner) class ratings.

For example, the author has visited a class 10 clean room which had an enclosed robotic process for photolithography. Inside the photolithography enclosure, the class was 0.1 (no more than 1 particle per 10 ft³.) No human entered this chamber; a robotic arm moved the silicon wafers from station to station, keeping track of each one individually. When 25 of the wafers were completed, they were moved to a small window where they could be removed by a fully-clean-room-dressed employee and taken to the next process.

Human workers are the dirtiest thing in clean rooms. The equipment can be super-cleaned after installation; the room itself is super-cleaned after construction. The filters are installed, the air begins to move, and before any production can take place, the clean room has to be tested to the desired level of cleanliness. Once this is accomplished, it is fairly easy to keep clean, except for the people that keep coming and going. Each time a person comes in to start a work shift, they bring with them the dust, pollen, lint, etc., from the outside world. If allowed to enter a cleanroom with all these contaminants, it would be impossible to keep the cleanroom clean.

To solve this problem, workers first dress in clean, lint-free clothing, covering them from head to toe. Gloves are also worn, as is a face covering and goggles or safety glasses. There isn't a bit of exposed skin, because skin is constantly shedding the outer layer of dead skin cells. No makeup, perfume, or cologne is allowed on workers, because it outgases (gives off contaminating gases). Then workers walk through an air shower, which shoots clean air on them as they rotate, knocking off any particles which may have attached themselves to their cleanroom garb back in the change room.

In the end, IC cleanrooms still struggle to stay clean, and are constantly studying the particles that interfere with the process, where they come from, and how to get rid of them. It is an ongoing process that becomes more difficult each time we shrink the size of the features on the IC. Particles too small to matter at a resolution of 0.15 µm suddenly become show-stoppers at a resolution of 0.05 µm (50 nm), so each time we improve (shrink) our resolution, we then have to go after removing even smaller particles.

Yield

Yield is an interesting concept that in most industries is not even measured. Yield is very simply the number of good products made, divided by the total number of products made. For example, if a company makes 1000 boom boxes a day, and 990 of those boom boxes are good, they would have a yield of 990/1000, or 99%. But companies that make boom boxes don't consider this very good; most have a yield of 100%, which is why it's not measured. The same is true in most industries, because if a product has a problem somewhere along the process line, the product is fixed and it continues on in the process. In the end, essentially all of the products made are good. The only exceptions to this would be the very rare occasion when one is dropped and it breaks into a thousand pieces, or is dropped into a vat of paint or some other unpleasant chemical, or some other major disaster occurs to the product.

Imagine now a process that considers it *quite* good when they can make 80% of their products good! This means that they throw away 20 of every 100 they produce! Such is the IC industry. The process of making an IC is so difficult that yields of 80% are considered quite good! In fact, when a company is first making ICs with a new (smaller) resolution, it is not unusual for them to consider any yield greater than 0% something to get excited about!

Another example is when a company begins to produce a new, very complicated design, such as a new microprocessor or IC for a cell phone. Depending on how complicated and new the design is, first-time yields of 5% are wonderful news!

Part of the yield problem is that if any mistake is made during the IC manufacturing process—if any single transistor or resistor or capacitor doesn't work as it should—the entire IC must be scrapped. Now consider the challenge when there are between 200 and 500 different steps, taking from 2 to 5 months to complete, one after the other, and there are over 1 billion devices on the IC! Anything that goes wrong with any of these steps or devices will cause the IC to fail. And until about 20 years ago, almost no practical methods existed to fix any of the problems, and so the yield suffered.

Today there are several methods available to fix many of the problems that routinely occur, and they are widely used to improve yield. For example, in a 2-Gbit memory chip (which contains over 2 billion transistors and capacitors!), it is very rare for an IC to finish the entire manufacturing process with all 2 billion devices fully functional. If IC manufacturers had to depend on raw yield, it would be less than 0.001%! Thankfully, repair options exist, with which today's IC manufacturers can boost this yield up into the range between 40% and 80%, which becomes viable. But even at these yields, much product is discarded because it cannot be repaired. Such is the nature of the tremendous complexity involved in making ICs.

Silicon Wafers: Polished to a Mirror Finish

If you were to see the top face of a silicon wafer, you would have a difficult time distinguishing it from a mirror. Indeed, the author has on many occasions seen them used in this application: a silicon wafer, appropriately mounted, allows an engineer with her back to the hall, to see people as they enter her cubicle!

One often wonders why silicon wafers have such a smooth surface, and with good reason. Such surface uniformity (flatness and smoothness) is not natural or easy to come by. But it is necessary if we want to make a good IC. This is best explained by describing the concept of *depth of focus*.

Depth of focus limitations can easily be noted with your eyes. Hold your thumb up about 12 inches away from your face and focus on it. As you keep your focus on the

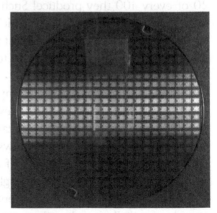

Figure 4-5:
An 8" (200 mm) silicon wafer with partially completed ICs.

thumb, use your peripheral vision to observe the area around you; notice how everything else is out of focus, unless it is also about 12 inches away. Now focus on something over 100 feet away; notice how everything around that same distance, give or take a few *dozen* feet, is also in focus. So in a nutshell, the closer something is as we focus on it, the narrower the range (depth) of things also within this focal range.

Now imagine trying to focus on something extremely close, with magnification. Magnification increases this problem, yet is essential to allow the image to focus exactly where we need it to. This means that if we set the focal distance to 2 mm, anything not exactly 2 mm away will be out of focus, or blurry. So, if we try to focus the light through our mask onto the photopolymer on the top of the silicon wafer, anything on the wafer that is out of this plane of focus (anything not flat or smooth) is not in focus, and will cause problems with the photolithography process. It is for this reason that silicon wafers are polished to a mirror finish.

Something else should be said here about silicon wafers. In the world we live in, it is extraordinarily difficult (read: impossible) to make anything absolutely pure. The purest gold is only pure to about 99.999%, which may sound pretty good, but would not be nearly good enough if we wanted to make ICs out of it. Over the past 50+ years, we have spent hundreds of billions of dollars and hundreds of millions of person-hours learning how to make silicon purer and more consistent. Any imperfections in the crystalline silicon structure will cause a transistor, resistor, or capacitor to fail. Given that there are hundreds of millions, or possibly billions, of transistors on many modern ICs, this means that the silicon must be purer than 99.99999999% (otherwise written as 100 parts-per-trillion, or ppt), or it will cause significant yield problems which cannot be repaired. Making silicon (or anything!) this pure is not easy to do, and it is only due to the fact that so much focus, time and money have been spent on this problem that we are able to make silicon purer than anything else on earth.

Testing: Is My Microprocessor Chip Really a Good One?

Testing is an essential function in all manufacturing. In the process of designing a product, such as a hair dryer, someone first establishes the basic functions the product needs to have. Such details include the weight, how many watts of heat it puts out, how much air the fan will blow, how long the cord will be, etc. As the design proceeds past the proverbial drawing board, those who plan to manufacture the product clamor for some way to assure that these functions are operational in each unit after it is manufactured. This assurance is the job of the testers. Sometimes testing is done at several stages along the way: test the fan, test the switch, test the heater, test the weight of each part. Sometimes testing is done only at the end: does the completed hair dryer meet all the specifications (does it do everything it is supposed to?)

For many products of today's manufacturing, the best way to assure that the final product meets all the specifications actually may not involve testing at all, but rather involves careful control of the process by which each part is made. If we know the process to make a perfect fan, and then carefully control this process as each fan is made, we can have a remarkably high degree of confidence that each fan will function as desired. If we do this for each part of the hair dryer, and also for the process by which the entire assembly is put together, we can have this same remarkably high degree of confidence that each completed hair dryer will function as desired. This method of reducing the need for testing is known as *process control*, and is widely used in the manufacturing industry today because it is so successful.

Process control depends very heavily on having processes that can be carefully monitored and controlled. This sounds redundant and painfully obvious, but the fact of the matter is that sometimes the degree of control necessary is not physically possible. This is very much the case in the IC manufacturing industry. Each of the processes used is very carefully monitored, and every parameter that can be controlled (such as temperature, time, and pressure) is controlled as well as we can control it. However, the control requirements for the many processes for manufacturing ICs are beyond the capabilities of the equipment, and each time we finally learn to make equipment that can do what we need, we shrink the transistors again and find that our equipment no longer meets our needs. Basically, we are constantly pushing the edge of what the equipment can do, which results in ICs having many potential problems, each of which can cause the ICs to fail.

Process control has been used very heavily in the IC manufacturing industry for many decades, and it has helped greatly in keeping yields

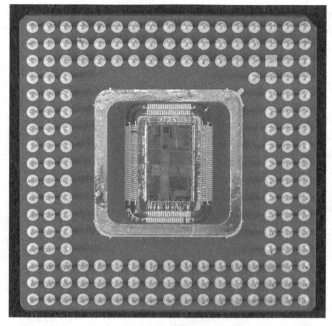

Figure 4-6:
A packaged IC. Visible is the actual "chip", as well as the tiny gold wires connecting it to the gray ceramic substrate, and thus to the pins.

high enough to keep the industry profitable. But process control alone is not sufficient to assure that each IC produced will function as specified. For this reason, each IC produced MUST be thoroughly tested. And this is no simple task!

The complexity of ICs has been increasing right along with Moore's law. This also means that the functions of ICs have been growing in number, making the testing challenge ever greater with each new generation of IC. For instance, consider the challenge of determining if each transistor, resistor and capacitor is fully operational in an SSI IC (with maybe 10 transistors). All one has to do is to determine how to test one transistor, then apply that test to each of the ten transistors. Because there are only 10 transistors, this is relatively simple.

Then consider the challenge when there are millions or billions of transistors, and only a relatively small number of them are easily reached. This would be analogous to trying to take a poll in a very large city, but not a statistically random poll. This poll must include every person in the city, but the only people you can query are those on the outskirts of the city, and those that you can convince to come to the outskirts. For this poll, you must adopt a totally different strategy. You convince each person you meet on the outskirts to take your poll to several people deeper within the city. Then these people come back to you with the results of your poll. This must be done hundreds, even thousands of times, before your poll reaches each person in the city.

In an IC, only a small percentage of the transistors can be directly reached through the pins that electrically connect the IC to the tester (and later to the electronic product of which it will become a part). For example, in a typical IC there may be only 200 pins, but 100,000,000 transistors. So each pin must be used to indirectly access information about all the 100 million transistors, which is quite complicated and does take a lot of time. Any transistors that cannot be tested directly or indirectly are untestable transistors, and are a risk to the IC manufacturer. If our 100 million-transistor IC has 99% of its transistors testable (which is actually quite high by today's standards), this means that there are 1,000,000 transistors about which we know nothing! If our process is such that over 99.99% of all the transistors we make are good, this still means that there will be 100 of these 1,000,000 untested transistors that are bad (on average), and we can't tell that they're bad!

Unfortunately, just because we cannot tell if they're bad doesn't mean the customer won't eventually be able to tell. The challenge is a classical tradeoff. We could spend more time testing our 100 million-transistor IC, and thus reduce the number of untested transistors, but this becomes quite time-consuming, and time is money in manufacturing. Shipping these 100-million-transistor ICs with many of their transistors untested *is* a risk, but a relatively small one. In the end, each company must decide what the acceptable *shipped-product quality level* (SPQL) is for each of their ICs, and adjust their testing accordingly.

What does this mean about that amazingly complex microprocessor chip in your computer? (A typical microprocessor will have over 300 million transistors and about 400 pins). It means that the manufacturer doesn't really *know* if it works completely, but it does have a very high degree of confidence that it works. It is possible that, in certain situations, over the thousands of hours it is used, the microprocessor will not perform as it should in very unique situations, but these situations will be rare, and are generally not greatly detrimental to the performance of the microprocessor. This uncertainty is one of the challenges we have come to accept, due the tremendous complexity of modern ICs.

What Can ICs Do?

ICs can assist in nearly any endeavor where information is processed. They can be used to sense the environment around us, including vibration, acceleration, humidity and temperature. They can perform mathematical calculations, repeatedly, and without error. They can store information. They can transmit and receive complicated communications signals. They can amplify signals, and even reduce the noise and the errors in some signals. They can play back or record music. They can detect enemy signals, equipment, or movements. They can help us see in the dark or see through the human body.

This list could go on a very long time. Perhaps a better question is, what *can't* they do? They can't tell the difference between a good decision and a poor decision, unless we previously endow them with a program to this effect. They can't exhibit true, unembodied intelligence (although they can sometimes appear intelligent!) They can't do anything they're not designed or programmed to do. They can't solve world hunger, or remove despots from power (although this could be argued!) They cannot think on their own. They cannot operate without electricity.

There are surely other things ICs cannot do, but which are not worth mentioning. The main point of this section is that ICs can be designed or programmed to assist in nearly every endeavor of human life (and some in our pets' lives!). They are incredibly versatile, limited only by imagination and cost, power and complexity. Without them, our world of electronics would have nothing more advanced than the products we had in about 1965. What a different world that would be!

So Much for So Little?

How can we really get so much function from an IC for so relatively little cost? After considering all the complicated processing that goes on in expensive clean rooms, populated by well-paid and highly-trained technicians, operators and engineers, and filled with very expensive equipment; and after requiring hundreds of steps over several months to complete an IC, how is it even possible that they can be produced for so little? It would seem that they would cost hundreds or even thousands of dollars each. But the average IC costs only a few dollars, which is truly an amazing accomplishment!

There are two main answers to this question. The first is automation; the second is volume. Automation has been heavily used in the IC industry for decades. The many steps required to make an IC must be done in exactly the right order, under exactly the right conditions. It has long been impractical to depend on human operators to be consistent enough to properly carry out the steps in the right order, and to control all the necessary processes. Considering the dire consequences of an out-of-sequence step, or of a process run under the incorrect conditions (*all* the product would be scrapped; few if any repairs are possible for these kinds of errors), the IC industry long ago turned to the consistency and dependability of automation to solve these problems.

For example, it is not unusual to tour an IC fabrication facility and wonder why nothing seems to be happening! The reason is that all the silicon wafers are in some expensive piece of equipment somewhere out of sight, having some automated process step performed on it in a special enclosure. Movement of the wafers between pieces of equipment takes only a few moments, while the process steps take much longer.

The volume of ICs is the other factor that helps bring the price down significantly. Most equipment can process several wafers at a time, and most wafers contain from 200 to over 1,000 IC chips (the actual silicon of an IC), so most processes operate on many hundreds or thousands of chips at a time. If the process must be done one chip at a time (as in the case of some photolithography), it only takes a few seconds, so one step on a wafer can still be completed in only a few minutes. A typical IC manufacturing facility will turn out tens of thousands of ICs every shift, and all IC manufacturing facilities operate 24 hours/day, seven days/week.

The steady march of Moore's law has led us all to believe that tomorrow's ICs will provide more function for less cost. This has held true for many decades, and the whole IC industry (as well as the general public) hopes it will continue to hold for many years to come.

Chapter Take-Aways

The amazing electronic products that we enjoy are only possible due to the invention of the IC and the subsequent continuation of Moore's "law" for the past several decades. Indeed, even the very manufacturing and testing of ICs requires very sophisticated electronic equipment, itself populated with many ICs. As is often said, a computer designer must have a computer to design one. And so with ICs: an IC cannot be designed and manufactured without many other ICs which have already been designed and built.

The IC industry, and the subsequent industry that manufactures electronic products that use them, is the largest single manufacturing sector in the world. There are hundreds of facilities ("fabs") that manufacture ICs; there are thousands of facilities that manufacture electronic products. And the volumes of these sold every year are truly staggering.

The average car has over 20 embedded computers (which means they have no keyboard or monitor, but are there keeping things going generally without our knowing it). The average cell phone has billions of transistors. The average portable computer has many tens of billions of transistors. Even a device as simple as an MP3 player costing only $30-$50 will have several billion transistors. A simple memory stick (aka: flash drive, jump drive, USB drive, etc.) with 16GB of storage has over 128 billion transistors.

Yet for all the incredible complexity that exists in these amazing electronic devices, we have grown accustomed to the idea that in three years, every one of these will be obsolete, and can be replaced with cheaper units that have more function. We have grown accustomed to this because it has happened for decades, but we should recognize that NO other industry in the history of the world has ever been able to accomplish this feat of continually dropping the price while improving function. And it is the IC and Moore's "law" that has made all this possible.

Conclusion

Earlier in this chapter, the improvements in the IC were compared to the automobile. Below is a table summarizing the data that went into this comparison. As laughable as the comparison may be, it is my hope that this comparison makes the point that the progress that has occurred in electronics, due to the ever-shrinking integrated circuit, is truly a miracle.

	1960	2012	
		Using Communication Comparison	**Using Computing Comparison**
Speed	80 mph	320,000,000 mph / 10,000 mph (with wind resistance)	6.4 Gmph / 28,000 mph (with wind resistance)
Cost	$10/100 miles	$0.005 / 100 miles ($5,000 / 100 M miles)	1 n¢ / 100 mi. (10¢ / 100 M miles)
Efficiency	15 mpg	30,000 mpg	4.80 Gmpg
Weight	4000 lbs.	1 lb.	.004 lbs (~ 1/16 oz)
Life	80,000 miles	600 M miles	25.6 G miles
Purchase Cost	$2000	$0.75	$0.15

Table 1: Comparison of improvements in electronics to improvements in automobiles.

CHAPTER 5
ELECTRICITY, ELECTRONICS AND SAFETY:
ALWAYS DANGEROUS YET EVER SAFER

After the amazing advances outlined in the preceding chapters, and particularly the immediately preceding chapter, some of this chapter could seem somewhat pedantic. But to put it in the proper light, one must have an appreciation for the power behind electricity.

All sources of power have the potential to injure us if improperly used. The power of steam preceded the power of electricity, and was accompanied by the dangers of steam burns, boiler explosions, and dramatic property damage. Burning coal or natural gas also has significant power, but has great inherent danger when the normally controlled burning process goes out of control.

So it is with electricity. When properly controlled and harnessed, it can be a great source of power for good purposes. But when any relevant safety procedure is violated, the consequences are terrible. Each year, approximately 300 people are killed in the United States by accidental exposure to electricity; an additional 4,000 to 5,000 people are injured enough to require time off from work. Clearly, the danger is very real, and electricity must be respected. Thankfully, many things have been done to improve the overall safety of electricity. As is the case with all sources of power, the greatest danger is an uninformed public.

Basic Safety Precautions

One of the first basic safety precautions for electricity was the design of the household outlet. Safety is the reason that the household outlet contains the receptacle (female) end of the connection, and not the plug (male) end. Imagine the danger if the household outlet contained the plug end—how much easier it would be to accidentally come in contact with live electricity with prongs sticking out of the wall!

Some of the basic rules with electricity include:

1. Always make sure something electrical is unplugged or switched off before working on it;
2. Make sure someone can't accidentally plug it in or switch on something while you're working on it;
3. Always wear good shoes while working on electricity;
4. NEVER mix water and electricity.

To understand the science behind this, let's first understand some things about electricity and about the human body.

There are two types of dangers posed by electricity: direct and indirect. Indirect dangers include explosions, falls, and reactive injuries. In very flammable or volatile environments, electrical sparks can cause an explosion. If a person is on a ladder, roof, or other elevated object, there

Figure 5-1:
A household plug and outlet. Note that the prongs are on the normally unenergized (plug) end, NOT on the normally energized (outlet) end.

is the possibility that accidental contact with electricity could cause them to fall. And reactive injuries result when we instinctively jerk our hand or leg back when unexpectedly coming in contact with electricity—this sudden motion can cause our hand, arm, foot or leg to strike a sharp object which could be nearby. Indirect dangers are a very large part of the overall danger that electricity poses, but are also the easiest to understand, since the outward manifestations of the injury are so obvious.

The direct danger of electricity is simply *electrocution*, which is death by electric shock. However, the process by which this can occur to the human body is not as simple as one might think. The degree of electric shock one experiences is a function of the amount of *current* that flows through their body, and the duration of *time* that the current flows. Currents less than about 5 milliAmps (thousandths of an Amp, or simply mA) are not sensed by the body in normal conditions and are therefore harmless. From about 10 milliAmps (mA) to about 30 mA the body will register the sensation, but the danger is minimal. From about 40 to 60 mA, the body will respond in several negative ways, depending on what part of the body comes into contact with electricity; these include a "can't-let-go" response by the hands, a violent shaking of the legs or arms, or possible fibrillation of the heart. Of these responses, the most dangerous by far is coronary fibrillation; more on this in a bit.

As the current flowing through the body increases, the danger increases. From about 70 mA to 100 mA, there is sure to be some kind of lasting damage to the body, especially if the current flows through the chest area. Above 100 mA can cause burns; above 10 A can cause dismemberment. While these currents are very dangerous, recovery is usually possible if the current is below about 150 mA and competent medical help is readily available.

Coronary fibrillation is, by far, the most frequent lethal response of the body to electricity. The human body activates the muscles with small electrical signals, as shown by Galvani many years ago (see Introduction). If another stronger electrical signal interrupts these electrical signals of the body, the mind loses its ability to control the muscles because its small electrical signals are overridden. Such is true with the heart, the most important muscle in the body.

The heart receives regular signals from the brain, telling it to contract in its rhythmic pattern. The electrical distribution system of the heart itself then spreads this message and causes the heart to contract. If an electric shock interrupts this regular signal, the heart doesn't know what to do and goes into *fibrillation*, which is when the heart simply quivers, instead of beating regularly. In this condition of fibrillation, no blood is being pumped through the body, and the victim will die in minutes. The purpose of *defibrillators* is to apply a dose of electricity large enough to cause the heart to reset itself and resume its regular beating. If the heart muscle is still in good shape, and if the victim has not been unconscious long, it usually works.

The success of the defibrillator, as used by emergency medical technicians, has given us another marvel of modern electronics: implantable pacemakers and defibrillators. The implantable pacemaker is useful in patients whose electrical signals to the heart are irregular or too weak, and their heart experiences *arrhythmia* (a beating pattern which is not regular or rhythmic), or their heart beats too fast (*tachycardia*) or too slow (*bradycardia*). Such patients are greatly aided by pacemakers, which produce their own regular electrical signals and help regulate the beating of the heart.

Some patients can experience an even more serious type of beating abnormality, in which the heart, on its own, goes into fibrillation. Unless medical help is available immediately, they will die in minutes. Modern medicine has advanced to the point that doctors can often tell what patients are at risk for this problem. In these cases, they can have implanted a pacemaker which can also defibrillate automatically when necessary. These *implantable cardioverter defibrillators* (ICDs) use electronics to monitor the beating of the heart, and send corrective signals to the heart as necessary. Since their invention, pacemakers and ICDs have helped millions of people both extend their lives and improve the quality of their lives. It is amazing to think that millions of people today walk around with life-saving electronics right inside their body!

Why Three Wires?: The Addition of the Ground Wire

An electric circuit, as in a flashlight, needs only two terminals or wires to operate. The positive lead of a battery can be thought of as supplying the current, which flows through the load (the bulb of the flashlight) and then goes to the negative lead of the battery. In the United States, electricity for the house is supplied at 120 Volts of alternating current (AC) at 60 Hertz (cycles per second). This means that the voltage is first positive, then negative, and that this cycle repeats 60 times per second, with the current flowing back and forth. But it doesn't matter if the voltage is supplied as direct current (DC), as in batteries, or as AC; it still only takes 2 wires for the circuit to operate. So why, in about the 1940s, did the electric industry move from 2 wires for electricity to the home or the factory to the present 3 wires? The answer is safety.

In a properly wired house, the black wire in house wiring is the hot wire, and is the only one that can hurt you if you accidentally touch it. The white wire only conducts the return current, and has a very low voltage (if any). Electricity which energizes a load (such as our refrigerator) has current which flows through both the black and the white wire, but only the black wire has an applied voltage. This can perhaps be better understood by using the water metaphor again.

All the water that flows over a water wheel goes from the top of the water wheel to the bottom of the water wheel. The power in the water wheel comes from the difference in potential between the water being at the top of the water wheel and the water being at the bottom. The same is true of electricity. Electricity entering the load has a high voltage (potential); electricity leaving the load has no voltage (or very little). Voltage is pressure or potential; it pushes the current and makes it want to move. Once the current moves through the motor in our refrigerator, it has lost all its voltage, and returns to the outlet.

So, in theory, the white wire, which carries the current back to the outlet, would have no voltage and would therefore be safe. So, we could use this white wire to ground the refrigerator itself, assuring that the metal of the refrigerator could never have a dangerous voltage and thus shock someone coming in for a midnight snack. That is the theory, and it is sound theory. The problem comes in the practice.

In practice, when wiring a home or workplace, the connection points (where the wires connect to each other) are less than perfect, and sometimes can be quite poor. If a white wire were to have a poor connection back at the outlet, it is possible that the current flowing through it from the load would cause a voltage to be present and that this voltage would then also be present on the metal of the refrigerator. This dangerous condition occurred often enough that the third wire system was proposed.

In the three-wire system, the black and white wires perform their usual function, with the black supplying the current and the voltage, and the white returning the current without the voltage. The third wire is the ground wire, and it does not conduct any of the current. This means that, even if there is a faulty connection on the ground wire back at the outlet, the ground wire will still not have any voltage on it, so the metal on the refrigerator could never have a dangerous voltage present. This system works very well, and unless an outlet is incorrectly wired, today's appliances are very safe.

Fuses and Breakers

Fault conditions can exist in all power systems; electricity is no different. Besides the danger of electrocution, another danger of electricity is a short circuit. As discussed above, electricity is supposed to flow from the source, through the load (a refrigerator, for example), and back to the source. The load itself is what limits the amount of current flowing. For example, in the United States all house appliances use either 120 Volts (the most common) or 240 Volts (usually limited to high-power appliances such as air conditioners, water heaters, and kitchen stoves.) Yet even among 120-Volt appliances, there are very different amounts of current flowing. A hair dryer on the 1200 Watts setting will use 10 Amps; a lamp with a 60-Watt bulb will use 0.5 Amps; a clock radio (with the radio off) will use only 0.005 Amps. They all use 120 volts, but the amount of current varies widely, depending on the load.

A short circuit is a fault condition in which the electricity flows from the source directly back to the source, without going through the load. This means there is NO LOAD to limit the current. In such a condition, the amount of current that flows can be huge: 2,000 Amps or even more! Such a large amount of current causes an arc where the short occurs; this arc can easily reach temperatures over 4,000°F very quickly, which is more than enough to ignite anything flammable which may be near.

The purpose of fuses, which were invented very early after electricity began to be widely used, is to interrupt this fault current quickly, so as to limit the potential for damage. If a fuse is appropriately sized, it can be very effective. For example, if our refrigerator (which uses about 3 Amps) is fused by a 10-Amp fast fuse, a short circuit would cause the fuse to burn out in less than 1/10th of a second, which would immediately quench the arc and therefore limit the potential damage.

The problem with fuses is that, once they burn out, they must be replaced, which can be a bit problematic. The homeowner must stock the proper types of fuses, and replacing them is not always easy. This problem has been largely solved with the use of breakers in the home. A breaker is a specialized type of switch and it operates much like a fuse; it monitors the current flowing through the circuit, and if the current exceeds a certain amount, the breaker will open and the circuit will be de-energized. Basically, the main difference from a fuse is that a breaker can be reset and does not need to be replaced. After the cause of the short circuit is removed, the breaker can easily be reset and no fuse need be stocked.

The disadvantage of breakers is that they are larger and more expensive than fuses. These are the main reasons that most automobiles continue to use fuses instead of breakers to protect the many circuits they contain. The good news is that, over the decades, billions of dollars of damage has surely been avoided due to the use of fuses and breakers.

Best Yet: The GFCI Outlet

It has long been understood that many lives could be saved if only electricity could be quickly interrupted when it contacts a person. But fuses and breakers have been very ineffective at this, as they are not designed for this problem.

One of the key factors in determining the seriousness of electrical shock is the *duration*—how long the person was in contact with electricity. If the contact is very brief, the danger is very minimal. For example, a person sliding out of their car and thereby getting all charged up with static electricity will easily carry over 20,000 Volts on their person.

Figure 5-2:
A GFCI outlet on the left, and a normal outlet on the right. Note the Test and Reset buttons on the GFCI outlet.

When they touch something metal (such as the car door), the result-ing electrostatic discharge can easily be over 70 mA; one would think that 70 mA at 20,000 Volts would be lethal! It would, except that the duration of the shock is so small that very little danger actually exists.

If the duration of the electric shock is very short (less than 1/10th of a second), we could generally rest assured that the experience, while possibly unpleasant, would generally not be deadly. The purpose of a GFCI (Ground Fault Circuit Interrupter) is to accomplish this. But how does the GFCI outlet know when the electricity is going where it should (to your hair dryer, for example) and when it is going where it should not (through you!)? Actually, a GFCI outlet cannot tell this *exactly*, but what it CAN tell is that some of the current going out from the outlet is not coming *back*.

As discussed before, the voltage drops from 120 Volts to 0 Volts as the current goes through an appliance. However, ALL the current in an electrical circuit should flow from the outlet, through the load (your hair dryer) and back into the outlet. If ANY of the current does not make it back to the outlet, it is going somewhere it should *not* go. By carefully monitoring the outgoing current, and comparing it to the returning current, a GFCI can quickly tell that something is wrong. It can interrupt the electricity in about 25 thousandths of a sec-ond, more than fast enough to prevent a lethal shock.

A careful look inside one of these GFCI outlets discloses a remarkably simple sensor mechanism, based solely on the current flowing in and out of the outlet. All current flowing in a wire brings with it an associated magnetic field, caused by the current. This very fundamental property of electricity has been known for decades, and is unchangeable. Additionally, the polarization

Figure 5-3:
A GFCI outlet (top), compared to an ordinary outlet (bottom). Not only are GFCI outlets bigger—they also cost much more (about 30 times more)—but they are well worth it!

(which direction the north and south poles point) and the strength of the magnetic field is determined by the direction and magnitude of the current. If one were to sense the magnetic field of the incoming current, and subtract it from the magnetic field of the returning current, the sum would be zero, since they are opposite in direction. A very simple loop around the two wires (black and white) performs this function (sums the magnetic fields of the two current directions); if this loop produces a voltage above a certain threshold, it is due to a mismatch between the incoming and returning current, which means another current path (a *fault* path) has been created, which could very well be someone being shocked. The small current difference trips an electronic circuit breaker, which interrupts the electricity.

A GFCI outlet is now required within six feet of anywhere there is a water faucet, or where there is a bare cement floor. These modern marvels of electronics, costing a relatively few dollars, have saved tens of thousands of lives since their invention and widespread installation over 20 years ago.

Electronics and Automotive Safety

A CUSHION OF SAFETY: AUTOMOTIVE AIRBAGS

The concept of airbags is several decades old. But like many ideas that are conceptually sound, the widespread adoption had to await a practical implementation. When first conceived, airbags would have added several thousand dollars to the cost of an automobile. While it was generally agreed that they could save lives and reduce injuries, not even for the most expensive cars was it deemed practical, and so the idea languished.

The heart of an airbag system is the inflator; the brain of the system is the accelerometer and associated electronics. The inflator is much like the firing chamber of a rifle bullet; it has a primer and a main chamber. The primer in a rifle bullet is ignited by the compressive energy of the firing pin of the rifle; the primer in an airbag inflator is ignited electrically, much as the small rocket engines of model rockets. The function of the primer is to burn quickly and with enough energy to ignite the main chamber chemical.

In a rifle bullet, the main chamber contains the gunpowder, which burns very quickly, causing explosive gases to push out the projectile portion of the bullet. In an airbag inflator, the main chamber contains a chemical such as sodium azide (NaN_3), which also burns very quickly and produces rapidly expanding gases which then inflate the airbag.

The process for making the primer (usually called the igniter) and the main chamber for airbags has been around for several decades, and is closely related to the process used for making rifle bullets. The languishing of the idea for airbags was not due to a lack of technology for these parts. Instead, it was due to a practical solution for the brain of the system: the accelerometer and the associated electronics.

Over 20 years ago, very few simple solutions existed for measuring acceleration. Most accelerometers were easily $500 and up; more if multiple directions of acceleration (or *de*celeration, in this case) must be sensed. Additionally, precision electronics were necessary to take the electrical output of the accelerometer and convert it to a useful signal. The most important output of the brain of the airbag system is a very simple signal: "Inflate the Airbag". If this signal is not active, the airbag does not inflate; if the signal goes active, the airbag *must* inflate. Obviously, it is very important that this signal be reliable in *all* conditions. This is *much* harder to do than can be readily appreciated.

An ideal accelerometer would sense (respond to) only acceleration or deceleration. This seems painfully obvious, but in practice, most accelerometers also respond to things like temperature, humidity, and particularly vibration. The fault conditions of an airbag are particularly bad; one is that the airbag inflates when there is *no* accident (scaring the driver terribly), and the other is that the airbag does not inflate when there IS an accident. Either is potentially lethal, so it is critical to reduce the possibility of these errors to near zero.

"Why not reduce the possibility of these errors all the way to zero?", one might ask. The answer to this question lies in the sometimes frustratingly random world we live in. For instance, fix a pellet gun firmly in place, point at a target, which is also firmly fixed in place. Fire five pellets at the target and observe that there are five holes, not one hole. Why is there variation? Or observe the burning of a fire within a closed fireplace (behind glass doors); why do the flames move about and display random behavior when there is no breeze? Variation is a fundamental law in the world we live in; nothing is ever *exactly* the same as something else very similar to it.

Likewise, make 1000 accelerometers by the same, carefully controlled process, and you will get 1000 slightly different behaviors from these accelerometers. If the process by which they are made is very tightly controlled, they will be quite similar, but they cannot be *exactly* the same. So how do you make 100 million accelerometers and have them *all* behave close enough to the same so that you can put them in 100 million cars and have them inflate the airbags *only* under the exact conditions they were designed to? Until about 15 years ago, the answer was that you just didn't. Each accelerometer used in industry (and there were *many*) was carefully characterized, calibrated, installed, and tested, a procedure prohibitively expensive for the consumer automobile market.

The brain of today's airbag systems uses accelerometers and the necessary electronics all integrated together on a single chip! This allows these to be mass produced, and testing and proven procedures learned in the IC industry have been put to work to allow these to be individually characterized, calibrated, and tested in large quantities. Finally, the electronics could be built on the same piece of silicon where the accelerometer was made, which brought the entire mechanism down to an affordable range, and with a dependable signal to inflate the airbag. Initially these fully integrated accelerometers and airbag systems were still quite expensive, adding just over $1,000 to the cost of a new car, so they were limited to the more expensive models, and only to the driver's side of the vehicle. Today, the cost has come down so much that two airbags (driver's side and front passenger's side) are usually standard, and multiple side-impact airbags are available in some models. Sure, we have learned to reduce the cost of making the inflators, but that alone cannot account for the more than 20-times reduction in the cost of airbag systems. And anyone who has had an airbag system deploy when involved in an accident knows of their general effectiveness in reducing passenger injury.

For just a few moments, let's take a look inside the electronics of an airbag system. The sensor is usually a small cantilever, which is basically like a very tiny diving board over a small (but empty) swimming pool. As the swimming pool moves suddenly in the upward or downward direction (in response to upward or downward acceleration), the tiny diving board (cantilever) bends. And just how tiny is this cantilever?

Typically about 1/100 the thickness of human hair in width! The tiny deflection due to acceleration causes a small change in the electrical properties of both the cantilever and the swimming pool; either can be used to give an electrical signal which is proportional to the acceleration. This signal is then amplified by the electronics on the same piece of silicon; after amplification, it is conditioned (noise is removed) and calibrated. If the final signal exceeds a given threshold, the electronics sends out its most important signal: a small voltage that tells the airbag inflator to trigger, causing the airbag to inflate. One of the very nice things about doing this all in electronics is how quickly it can happen. Mechanical things (such as the sudden acceleration) take several thousandths of a second (milliseconds), which seems fast to us, since we can do very little in such a small amount of time (not even blink our eyes!) Yet to the electronics, this is very slow. Electronic signals can be processed in *billionths* of a second (nanoseconds), which is about a million times faster than the mechanical things occur. So, most of the time, the electronics simply waits, very patiently, for something to happen. As soon as something does begin to happen, the electronics is on top of it, responding hundreds of thousands of times faster than necessary. The delay from the sudden change in acceleration to when the electronics sends out the signal to inflate the airbag could be as little as 100 nanoseconds; the time it takes the airbag to inflate is about 10 milliseconds!

No new technology is without its drawbacks, and airbags are no exception. Their major drawback has been the injury inflicted by them when they inflate in the presence of a baby or a small child. Because of the very short amount of time available for an airbag to inflate (only a few thousandths of a second), and the forces they must absorb (the impact force of a human body traveling up to 70 mph, or about 102 feet per second), airbags must inflate with substantial force. Small children or babies have been killed by this very substantial force, which has necessitated the guidelines that such children may not ride in the front seat of an automobile. But help is on the way!

Wouldn't it be nice if airbag systems were more intelligent? It would be great if they could sense if someone is in the passenger seat; if there is no one there, they need not inflate! It would also be great if they could sense the size of the person in the passenger seat; if it is a baby or small child, again it would not inflate! This would allow children to

safely occupy the front seat again, which would be much more convenient for the driver! Such systems are presently being introduced.
They are equipped with sensors (again, the kind enabled by advances
in microelectronics) that determine the weight of the passenger seat
occupant; if that occupant is too small, the airbag will not inflate.

And this is not all—there is even more help on the way! Wouldn't it be
nice if airbag systems were so intelligent that they would inflate the airbag at a rate proportional to the size of the seat occupant? Large occupants need a fully-inflated airbag to minimize injury; smaller occupants
need only a mostly-inflated airbag; the smallest occupants need no airbag inflation at all. Electronics exist that can determine this information,
and send it to the newest adjustable-rate airbag inflators. Now essentially
anyone of any size can occupy the driver's seat or the passenger seat,
and not be severely injured by the airbag itself! It's hard to think how it
could be better (other than completely avoiding the accident!)

ANTI-LOCK BRAKES

Another great invention that has greatly improved automotive
safety, and which has been enabled in large part by electronics, is anti-
lock brake systems, commonly referred to as ABS. The idea is simple,
and has long been known to professional drivers.

Maximum braking deceleration occurs just *before* the tires lose their
grip on the roadway. A car skidding will take much longer to stop
than one braking hard but not quite skidding. Additionally (and more
importantly), a car which is skidding is no longer under the control of
the driver, but is completely at the mercy of the road. A car which is
braking hard, but not quite skidding, can still be steered, and is thus still
under control of the driver.

But only highly skilled drivers learn to exert such careful control in emergency situations. The rest of us, faced with the need to stop very quickly,
instinctively apply the brakes to the maximum, which almost inevitably
causes the car to skid. A frequent pumping type of action on the brake
pedal can also prevent skidding; each time the brakes are released in the
pumping action, traction is restored and the car remains under control.
But learning and applying this skill in event of an emergency is even more
difficult than braking hard without losing control. So, why not let the car
do it for you automatically? This is what ABS systems do.

Again, the idea has been around for several decades, but the technology to apply it on a commercial scale has simply not been affordable. Brake systems which can be modulated (changed) under control of something other than the driver have been available for some time, but have not been affordable. Major advances in the manufacturing technology behind these systems brought their cost down by nearly an order of magnitude (10 times)—which is a lot, but was not enough. Again, the major cost decrease was in the electronics. Electronics that once took up several large boards and cost over $1,000 now can complete the same function, and do it even better, at a fraction of the cost (less than $50) and in much less space, using much less power. Now what was once affordable for only a few is available on many models of automobiles as standard equipment.

And the benefits are well documented! Insurance companies give discounts if your automobile has an ABS. Next time you are in an open parking lot (where your driving won't impact anyone else), just try to make your ABS-equipped automobile go into a skid—it is nearly impossible! It is truly a wonderful feeling to know that in the panic of an emergency, your car will remain in control, and will stop faster than you could have stopped it by fully applying the brakes.

Chapter Take-Aways

Modern civilization has generally learned to live with many types of power, including water, steam, and electricity. Of these, the power of electricity has become nearly ubiquitous. It is more widely available today than ever before, yet fewer people are killed by it than ever before. This wonderful improvement in safety would not be possible without much effort and without the major safety advances of a third prong, fuses and breakers, and GFCI outlets. While electrical faults do still occasionally start fires, this is still much rarer than in the days of candles and open fireplaces. We have made electricity much safer than in decades past.

Two of the most significant advances in automobile safety, airbags and anti-lock braking systems, have been dramatically enabled by advances in electronics, in addition to many other processing and manufacturing advances. These advances have greatly improved the safety of automobiles, much to the delight of everyone.

Chapter 6
Voice Is Nice, But . . . : Television

To truly appreciate the miracle of modern television, one must first understand how radio works, since the same principles are used for both, and it is much easier to start with something simpler. It is amazing to think that invisible, imperceptible signals surround us nearly all the time, and yet these very faint signals can carry both voice and video information. And the more one understands how this is done, the more amazing it becomes!

In Chapter 1: A Rudimentary History, there is a small section devoted to the development of radio (or "wireless", as it was then called). Picking up where that section left off, we have Guglielmo Marconi successfully sending the letter "s" (via Morse code) across the Atlantic Ocean. Due to the ease of transmitting and the very substantial range it provided, Morse code radio was used for many decades afterward. But listening to Morse code was not of interest to the general public, and with good reason!

Morse code was basically digital; the presence and duration of a pulse determined the characters being sent; it took much practice to be able to send and receive in Morse code. Ideally, one would like to pick up the transmitter of a wireless and simply talk to someone else. Voice is quite another matter, compared to the simpler digital Morse code. In order to make voice digital, it required electronic circuits which would not be available for decades, and not practical for decades more. The solution was to use voice to directly modulate (change) the electronic signal being transmitted.

Needed: A Carrier

Sending a message from one place or person to another always requires some kind of carrier. If sending a message via semaphore (flags), the carrier is light (the flags must be visible). If sending a message via flashlights, the carrier is also light, which is turned on and off (modulated) in a prescribed method to represent the message. And not surprisingly, if sending a message via carrier pigeon, the carrier *is* the pigeon.

Morse code is transmitted on radio signals in a way very similar to sending a message via flashlights. A sensitive radio detector at the receiving end listens for the transmission at a prescribed time and frequency. The transmitter simply turns on and off the radio signal in the method described by the code: a long pulse for the letter "t", a short pulse for the letter "e", and so on. These pulses travel up through the antenna and out into the air, where they travel literally at the speed of light (radio waves are the same as light, just lower in frequency) and are picked up at radio receivers.

What is the nature of the radio carrier? It is simply an electromagnetic wave in the same shape as the waves made by a rock splashing in a still pond (known as *sine* waves). These waves are not like those created by a single rock splashing in the pond, but rather like those created by a hand constantly pushing up and down on a board which is floating in the water. If the hand suddenly stops, the waves subside; when the hand starts again, the waves begin again. The electronic hand which does this is a circuit known as an *oscillator*, producing a constant electromagnetic sine wave (the carrier), rippling out through the air in an invisible manner, but detectable by radio receivers.

Modulating this carrier can be as simple as sending it to a small shutter, which blocks the carrier when closed and passes it when opened. Opening and closing this shutter would allow us to send Morse code via radio waves, much like turning on and off the flashlight would allow us to send Morse code via light. But opening and closing this shutter manually is not only tiring, but dreadfully slow; only a few pulses a second can be sent. This is where electronics again provides the solution, in the form of a switch. Putting the carrier through an electronic switch allows us to turn the carrier on and off without tiring, and it allows us to turn it on and off thousands, millions, or even billions of times a second, indefinitely. This gives us the advantage of being able to send more data in less time, which is a *very* significant advantage.

The above has allowed us to send Morse code *very fast*, but it is still only Morse code—not very useful to the general public. But what if there were a way to change the nature of the electronic switch, so that instead of only turning on and off, it could also be turned to any level in between? If such a switch could be created (see Chapter 2 on the vacuum tube), it would allow us to use the human voice as a signal to the switch, turning the carrier on, off, and to every level in between, in response to the voice signal. Voice at maximum amplitude would give the full carrier; voice at minimum amplitude would give minimum carrier; no voice at all would produce no carrier.

The system just described is very similar to today's AM radio (AM stands for Amplitude Modulation). The voice or music on the transmitter end modulates (or changes) the amplitude of the carrier in response to the amplitude of the voice or music. At the receiver end, this variation in carrier amplitude is detected by the electronic receiver circuit, and is turned back into the original voice or music. The great part about doing it this way is that we can now transmit voice or music directly over the carrier. This was first done in 1906, and within a decade, commercial AM radio had begun.

The system just described is a type of *analog* radio: it allows us to transmit signals with continuously variable intensity, just as the human voice varies. This is just what is needed, for voice and music have an amplitude which is continuously variable, and analog radio provides a way to transmit voice and music. But the AM system described above is not the only way to transmit voice and music— FM can also be used. This is frequency modulation, which means changing (by only a very small amount) the actual frequency of the carrier (FM = Frequency Modulation). For example, in the commercial FM band, consider a radio station assigned to transmit at 89.5 MHz (MHz = MegaHertz, or mega cycles-per-second). Such a station could modulate its carrier up to 89.6 MHz and down to 89.4 MHz (a deviation of ±100 kHz). This modulation is continuous, and allows us to put our analog voice or music on the carrier and detect it on the other end with a circuit that produces a voltage proportional to the frequency changes.

FM transmitters and receivers are more complicated than AM transmitters and receivers, which is why AM preceded FM by about two decades. Nowadays, the circuitry required to receive *both* AM and FM can be fit on a small piece of silicon about the size of two grains of salt, so the added complexity of FM is no longer an issue.

73

But if AM works, why invent another method of modulation? The answer lies in the usual fact of life: no solution is ideal in all ways. There are a couple of drawbacks to AM, the details of which are beyond the scope of this chapter. FM overcomes those drawbacks, but at the same time has some drawbacks of its own, which AM does not. This is why, even in the very complicated radio world of today where we have many wireless devices in our homes (cell phones, remote controls, radios, wireless mouse & keyboards, pagers, television, etc.), the actual modulation methods used are a mixture of FM and AM, as well as some very interesting variations on these basic themes.

So this section leaves us understanding the basic technology behind how a carrier is modulated to contain voice or music, and how that carrier is then transmitted and received. But this does not explain how pictures can be sent through the air!

What Is a Picture, Really?

As perceived by the human eye, a picture is a very high-resolution collection of very small picture elements (PIXture ELements, or pixels). The human eye contains rods and cones, each of which can sense light. While there are many of them (about 120 million rods, and about 7 million cones), they are not infinite in number. This means that, in actuality, the pictures perceived by the human eye are made up of millions of very tiny pixels, which the mind blends together to form what we see as continuous pictures.

Going from this example, it is easy to see that if we could convert a picture to many pixels, and if each of those pixels could be transmitted over radio, and if we could reverse that conversion process on the receiver end, we would have television! While this is definitely overly simplified, it is still the basis of thought that brought television to pass. The missing pieces (something to convert the picture to pixels and the pixels to an electronic signal, on the transmitter end; and something to convert the electronic signal back to a picture on the receiver end) were worked on early in the 1900s. Some success was achieved on the receiver end with a device known as a cathode-ray tube, or CRT, which is only now nearing the end of its life many decades later.

The idea of a CRT is really quite simple, although the actual implementation of it is very complex. Most of us are familiar with

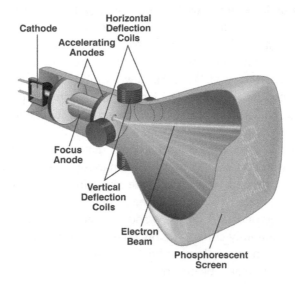

Cathode
Horizontal
Deflection
Coils
Accelerating
Anodes
Focus
Anode
Vertical
Deflection
Coils
Electron
Beam
Phosphorescent
Screen

Figure 6-1:
A cutaway diagram of a CRT, also showing a simple image being drawn on the phosphorescent face of the screen.

phosphorescent materials: those materials that sort of soak up light and then glow in the dark for a short time. Each time they are exposed to light, they glow for a short time afterward. Imagine, then, covering a wall in your home with this material, turning off all the lights, then waving a very bright flashlight across this wall briefly. Upon turning off the flashlight, one would see the path taken by the flashlight beam as it moved across the wall, although only until the phosphorescent material stopped glowing. Zorro could paint his famous "Z" on the wall with his light saber (OK, so we're mixing movies here, but the idea holds), and viewers would see it for a short time.

Now imagine if you could turn this flashlight on and off very quickly (hundreds of thousands of times per second), and if you could move it back and forth across the wall thousands of time a second. Using such a flashlight, start at the upper left corner of the wall, move it across to the right side, turn it off, move it back to the left side and slightly lower than the first time. Repeat (turn the flashlight on, move it across to the right side, turn it off, move it back to the left side, slightly lower than last time) until the entire wall has been exposed to this flashlight (the flashlight has moved to the bottom right corner of the wall). What you would see is the entire wall glowing. Now imagine doing this, but turning the flashlight on only when it is pointing at the middle of the wall as it sweeps across the wall. What the viewer would see is a single vertical line in the middle of the wall. Do this in a much more complicated way by turning the flashlight on and off many times as it crosses with each sweep, and you could actually paint a picture on the wall, although it would fade as soon as the phosphorescent material dimmed.

If such a process could be done continuously, and if the flashlight were turned on and off at exactly the same time each time it swept across the wall, the viewer would see a continuous picture, made up of bright pixels or dark pixels. Now add another control: imagine being able to continuously control the brightness of the flashlight, from full on to full off, and every level of brightness in between. With this control, we could paint a picture on the wall made up of bright pixels, dark pixels, and every level of brightness in between. This gives us a much more detailed picture, and much more like what we see in nature, but still only in one color (what we often call "black and white" or "monochrome").

Now Let's Add Motion

There are still two things lacking in this example: moving images, and full color. Let's consider first how to create images that move. To do this, we again need to understand something about how the human eye perceives the real world. There is an old saying among magicians: "The hand is quicker than the eye." Indeed, there are many things quicker than the eye. An example with which most of us are familiar are the wings of a hummingbird or a large insect such as a bumblebee. Their wings appear to be a blur to our eyes, partly because of something known as retention.

If the human eye is briefly shown an image, that image will be retained for about 40 thousandths of a second. This can easily be demonstrated by flipping through a deck of cards; flip through them very slowly and your eye sees each card individually. Increase the rate and soon they merge into each other, due to this retention. If each image is only slightly different than the previous image, the retention of the eye will cause them to look continuous. This is how motion pictures work. With film, there are 24 individual pictures per second, and since they differ very little from each other, the retention of the eye, together with the perception center of the brain, causes all these individual still images to be merged together in time into a smooth, continuous motion.

With our phosphorescent wall, the next thing we would have to add to get moving images would be a way to paint one picture in about 1/24th of a second, then paint a slightly different one in the next 1/24th of a second, and so on. After 24 of these, we would have a full second of what our eyes would perceive as continuous motion, right there on our wall! There is one other thing important to note here, and that is the retention of our phosphorescent paint. If the paint holds onto the light longer than about 1/24th of a second, the individual images will appear to smear instead of just move. Since this does not match what our eyes see in nature, this smear is undesirable.

Now let's take this wonderful phosphorescent wall onto which we have been painting moving pictures with our modulated flashlight, and let's put it inside our television set. The wall is the face of the television screen, but it is a transparent wall, made of glass. We will put the flashlight behind this wall, and the phosphorescent paint also behind the wall, to keep them out of our way. So, as the flashlight paints moving images on this screen, the light will be seen *through* the transparent glass wall, as we look at the other side of the wall. The flashlight in the case of a CRT is a beam of electrons, in the form of what we call a cathode ray (see Chapter 2 on the vacuum tube). This ray of electrons can be easily steered across the screen back and forth, up and down, and its intensity can be electronically modulated to any intensity, from full brightness to completely off, and everything in between. This beam goes from the upper left corner, across the screen to the right, is turned off, then brought back to the left side a little lower than before; this process is repeated 525 times ([1] see below), until the beam reaches the lower right corner, at which point the beam is turned off,

---

[1]The author has chosen to ignore interlacing, which is an artifact of the very old technology available when television was first invented. It was a compromise, and understanding it is not necessary to understanding the basics of how television works.

moved to the upper left corner, and the process repeats, 30 times a second, continuously. As these electrons strike the phosphorescent paint on the inside of the screen, we see the light coming out toward us, and what we see (when everything is working right) is a very nice, full-motion black-and-white picture.

It is marvelous that this beam of electrons can be steered relatively easily, and that it never tires of this continuous process of being turned on and off, moved up and down and back and forth, and being modulated in intensity. And just for fun, think about that beam; it sweeps back and forth 15,750 times a second, or about 56 million times for a 1-hour TV show! Although this process will not go on forever, it is not uncommon for it to last over 15 years when on 24 hours a day, 7 days a week, 365 days a year. That's a LOT of sweeps.

Now Let's Give It Color!

The light from the sun is the light to which our eyes are accustomed. To reproduce images that appear natural, ideally the light would be the same as the light from the sun. Unfortunately, that light is rather complex in nature, being made up of many frequencies (or colors), from infrared (which means below red—red is the lowest frequency of light we can see), through the visible colors of red, orange, yellow, green, blue, indigo, and violet (always in that order, at least in this universe), and into the ultraviolet (which means above violet—violet is the highest frequency of light we can see). Thankfully, there are some wonderful simplifications that we can use to produce very realistic pictures without all those frequencies of light; if this were not so, television would have been delayed by decades.

If we mix colors of pigments, such as coloring with crayons, we produce *subtractive* combinations, and if we combine all these colors, we get black, which is the absence of light (we have successively subtracted all colors). If we mix colors of light, we produce *additive* combinations, and if we combine all these colors, we get white, such as the color of the light we see from the sun. But there are three primary colors in light that, if mixed together in equal intensity, also produce light that our eyes perceive as white. Adjusting the relative intensity of these primary colors of red, green and blue allows us to produce all the other colors, and all their shades.

78

Figure 6-2:
The individual red, green and blue sub-pixels that make up every color pixel. Varying the intensity of each sub-pixel allows us to produce all the colors of the rainbow.

The phosphorescent screen we have previously described needs to be modified, and not too much, thankfully. Instead of only one color of phosphors, we now need three, so that each pixel is made up of red, green and blue phosphors. Now we also need three electron guns, one each for the phosphor colors, so that they can each be lit up independently.[2] With this setup, as we are all aware, we can indeed produce very realistic color images on our televisions.

There is one small catch here, however. Color television became possible in the late 1950s, after black-and-white television had already been around for several years. To successfully introduce color television to the public, when there were already hundreds of thousands of people using black-and-white TVs, we had to send a television signal that could produce black and white *and* color! So the decision was made and the technology put in place to allow broadcasters to send

---

[2]The author chooses to ignore the fact that some televisions actually use only one electron gun. This fact is not essential to understanding how color television works.

one signal (for each channel) which included both black-and-white signals and color signals. How this was done is really quite amazing, and is not too difficult to understand.

The intensity of each phosphor on a black-and-white screen is known as the *luminance*. The intensity of each phosphor on a color screen combines in what is known as the *chrominance*. It is not too difficult to send both the luminance and the chrominance in the same signal; this allows black-and-white TVs to use the luminance to provide a black-and-white picture, and color TVs to use the chrominance to provide a full color picture. So, since the early days of color television, this was the solution adopted in the United States in a standard known as NTSC (National Television Standards Committee). In other countries, which created their television systems later, other standards were developed, such as PAL (Phase Alternation Line rate) and SECAM (a French acronym, meaning roughly SEquential Color And Memory). Each of these standards allows a single signal to contain the information for both black-and-white and color televisions.

The Complexity of the Television Signal

One of the most amazing things about television is that it works at all, especially more than 60 years ago! The television signal itself is quite complex, being composed of several sections necessary to make the signal look good under as many conditions as possible. The NTSC standard (and the PAL and SECAM standards) are all analog standards, meaning that the signal can have an infinite number of values at any point in time. However, there are five very specific points in time concerning the video signal when very specific things are expected. These times occur once every frame; frames in television occur 30 times a second. The reason for the need of these times has to do with picture quality and picture synchronization.

Each TV set, if properly manufactured, should produce a picture which meets basic standards and is enjoyable to view. To do this, reference signals are included in the broadcast signal, so each TV set knows what white (the brightest the signal can be) and black (the darkest the signal can be) look like in the signal. These are the intensity reference signals.

Another part of the signal having to do with picture quality is the color burst. The purpose of this signal is to allow color TV sets to accurately reproduce black-and-white programming. If this color burst is present, the TV interprets the chrominance information to provide color. If this color burst is absent, it means that the signal is being broadcast as a black-and-white signal; the TV set interprets the absence of the color burst to mean it should *not* try to interpret the chrominance information, thus avoiding false colors in a black-and-white picture.

The fourth part of the signal, and which occupies the vast majority of the time of the signal, is the actual luminance and chrominance information for each of the 525 lines of NTSC television. This is the information that modulates the intensity of our cathode-ray flashlight, with which we paint the intensity of every pixel on the TV set, 30 times a second.

The fifth and final part of the television signal is the synchronization. The importance of synchronization is obvious to anyone who has ever watched a very old TV set in which the picture rolled up, or down, or sideways. If a TV set is not perfectly synchronized to the signal being transmitted, each frame of the moving picture will appear to start in a slightly different place, which makes it look like the picture is rolling in one direction or another. For the picture to look like it should, the cathode ray must start in the upper left corner at exactly the right time for each frame, 30 times a second. The horizontal and vertical sync pulses which are a part of the television signal take care of sending this information.

Broadcasters are all required to broadcast a signal which complies with the standard for the country in which they are broadcasting. If the TV set is working properly, it will interpret each of the five portions of the signal appropriately, and the picture will look great. If there is a problem with any of the many circuits, the picture will exhibit one or more rather sickly symptoms. The circuits inside the TV set must remain accurate and not drift with temperature, humidity, and age. This is not too difficult with today's technology, but with the technology available decades ago, it is truly amazing that these complex and precise functions could be implemented, and for reasonable prices.

And Finally, Let's Make it High-Definition TV!

It is truly amazing that as long ago as 1960, we were able to broadcast and receive a color television signal which has remained adequate for decades. Certainly the technology for the electronic circuits in a TV set has improved dramatically over that relatively long period of time, and the costs have dropped dramatically also. Yet in these several elapsed decades, until about the year 2000, no new standard had come along to give us a better television picture. It is worth digging into this a bit further to understand why.

The NTSC standard delivers 525 horizontal lines of resolution, with 384 pixels on each line. This is a total of 201,600 pixels (525 x 384). Each of these pixels must be updated at 30 times per second. This gives us a pixel painting rate of 6,048,000 pixels per second! It is not at all simple to transmit and receive this much information in only one second, and to do that continuously. And one thing it takes a lot of is bandwidth.

Bandwidth is very much like lanes on a highway. If a single lane can handle 20 cars a minute, two lanes can handle 40 cars a minute. If you need to handle 200 cars a minute, you need 10 lanes. Likewise, if you need to send 200 pixels per second, you need approximately 200 Hz of bandwidth. So, our 6,048,000 pixels per second requires approximately 6,000,000 Hz (6 MHz) of bandwidth! This wouldn't be a problem if there were an infinite amount of bandwidth available, but there is not. So, back when the original television standards were set, television was allocated 6 MHz for every television channel. For instance, channel 2 in television occupies the bandwidth from 54 MHz to 60 MHz; channel 3 occupies the next 6 MHz from 60 MHz to 66 MHz. These are wide bandwidths because the signal to be transmitted in that bandwidth must contain our 6,048,000 pixels each second! If our television had a much poorer picture, such as 300 horizontal lines and 200 pixels on each line, this would require only (300 x 200) = 60,000 pixels, and at 30 times per second, this would be only 1,800,000 pixels per second.

In electronics, there are *many* sizes of cars that we need to put on our highways. Some are only fractions of an inch wide; others are many feet wide. For example, if we consider the bandwidth needed to send a phone conversation and liken it to a highway lane ¼ inch wide, the bandwidth needed to send an AM radio station would be about one inch wide; the bandwidth needed to send an FM radio station would be about 17 inches wide (about 1½ feet); but the bandwidth needed to send a television station would be about **42 feet** wide! So providing the room for many television stations requires a very wide highway! Just the channels from two to thirteen would require a highway over 500 feet wide, and that doesn't allow for any of the UHF channels! Including all the UHF channels would require a highway over 2,800 feet wide (over ½-half mile!)

So, this is the bottom line for the reason why the television standard has remained unchanged for decades: there simply is not enough bandwidth available to improve the television picture, since an improved picture would require more pixels, which directly means more bandwidth!

So 20 years ago, when work on higher-resolution TV began, how could we even dream of doing high-definition television (HDTV) given these bandwidth limitations? There are only two ways to do this. The first is to change the allocations for all the existing television stations, which basically means tearing up our 2,800-foot wide highway which we built for television channels 2 through 69, and rebuilding the whole thing. And while it's being rebuilt, no one can use the highway, which means no television! A more practical way to do this would be to tear up and rebuild only a few lanes at a time, but that would take a very long time, and would still be a big mess. The only other way to do HDTV is to find some kind of magic way to shrink the bandwidth the signal would require. This is very akin to pushing an elephant through a straw! But the magicians of electronics are ever at the ready, and have indeed provided the solution, and it is not through smoke and mirrors, but through something known as compression—slimming the elephant down so much that he actually *does* fit through a straw!

So How DO You Stuff an Elephant Through a Straw?

The first thing one must do to allow for compression is to convert the analog television signal to digital. There are many advantages to doing this, and most of them will be covered in Chapter 8 on deep space probes. One of these advantages is that digital data can be compressed, while analog signals (in general) cannot. But converting the analog television signal to digital also creates a problem: suddenly we need MORE bandwidth than before! Why this is so will also be explained in Chapter 8, but suffice it to say that by converting our 6

MHz analog signal to a digital signal, we now need to transmit about 120 million (Mega, or M) bits per second, which typically requires more than three times our available 6 MHz bandwidth. By going to digital, we have tripled the size of the elephant!

And this was only the start! We also have decided to go high definition, which means more horizontal lines and more pixels per line. Specifically, we have chosen a picture with 720 horizontal lines and 1280 pixels per line, which gives 921,600 pixels, or nearly 1,000,000. At a resolution of 10 bits per pixel, and a rate of 30 screens per second (same as before), this equals 300 M bits per second! So, by going to HDTV from ordinary (NTSC) television, the elephant has grown another three times larger!

But this is where the magic of compression saves the day. There are many types of compression, and the details of how many of them work are beyond the scope of this discussion. But thankfully, two of the most useful and powerful types of compression are not difficult at all to understand. These compression types are *spatial* and *temporal* compression.

Spatial compression depends on the fact that adjacent pixels usually do not differ by very much. For example, if the picture on our television is of a Wimbledon tennis match, most of the screen is green, and there are many adjacent pixels which are green. Thus, to send the information of the color of the first horizontal line, we may only need to send the full color information for the first pixel, then send the very small differences between all the other green pixels on this line. This takes many fewer bits than it does to transmit the full color information for each pixel, one pixel at a time. The same is true for adjacent horizontal lines; once we have transmitted the first line, we can simply send the differences between it and the next line, and again we will save hundreds of bits. Using spatial compression alone, we can reduce the required bandwidth by several times.

Temporal compression depends on the fact that in the 1/30th of a second that elapses between picture frames, there generally is not a great deal of movement, and therefore not very much difference. So, after painting the first full frame of a picture, we need only transmit the differences between the next frame and the current one. This approach also allows us to reduce the required bandwidth by several times.

As mentioned before, both of these types of compression are possible only because we have converted our signal from analog to digital. Once the signal is digital, lots of amazing things can be done with it, and compression is one of the most useful things. In fact, digital compression is so effective that we can reduce the size of our elephant to LESS than the size of our straw (narrower than the 42-foot wide highway lane needed for one television station). Broadcasters love the new HDTV standard; even though it requires them to install completely new transmitters, it also gives them more room on the 2,800-foot wide highway, room which they can use to transmit other signals (and therefore charge other customers). By going digital and compressing it, we have actually REDUCED the required bandwidth for a signal which is digital AND higher-definition, each of which inherently requires MORE bandwidth.

So, how do you stuff an elephant through a straw? By using compression, which makes our elephant so thin it goes through the straw with room left over!

Chapter Take-Aways

I hope that after reading this chapter, or indeed even while reading it, the reader jumps up out of curiosity and obtains a magnifying glass to see the red, green and blue pixels on their television screen. But, if you don't have a magnifying glass, a simple drop of water on the screen will suffice—try it!

I also hope that the next time you watch TV, you think of the amazing complexity of which the signal is made, sending information on the color of each of nearly 1,000,000 pixels, 30 times per second, and so perfectly synchronized with the transmitter that the pictures are incredibly realistic!

CHAPTER 7
SEEING WHAT IS TOO FAR AWAY
TO SEE: RADAR

What Does It Mean to "See" Something?

The concept of vision is truly one of the most amazing capabilities of the human body. The light perceived by the human eye is made up of many colors or *frequencies*. The lowest frequency of light we can see is red in color; frequencies below this are *infra*red, or below red. The highest frequency of light we can see is violet in color; frequencies above this are *ultra*violet, or above violet. It is through the interaction of these various colors of light with the objects of our surroundings that our eyes perceive. But what we see is, by no means, all the light[1] there is.

Many of us are familiar with the fact that sunburn and rapid skin aging is not caused by the light we see, but rather by the ultraviolet light. Most of us are familiar with the huge amount of heat that pours into an automobile in the summer, far more than can be accounted for by the visible light; this is the infrared or heating frequencies of the sun. These infrared and ultraviolet frequencies are simply light at lower and higher frequencies, not visible to the human eye, but still very much real. Additionally, there are many frequencies of light beyond these! (See Figure 7-1).

Above ultraviolet light is another range of frequencies we know as x-rays; above these is another range we know as gamma rays; and cosmic rays are above that. Below infrared light is a range of frequencies we know as microwaves; below that we call them radio waves, which are further broken down into the ranges of UHF (ultra-high frequencies), VHF (very-high frequencies), and so on down. These radio frequencies include the familiar bands used for FM and AM

---

[1]Throughout this chapter, the word "light" will be used to describe what scientists and engineers call "electromagnetic radiation". While this is admittedly an oversimplification, it should suffice for our purposes in this discussion.

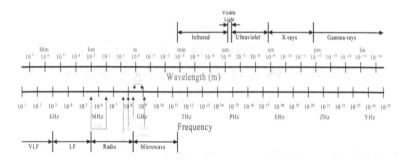

Figure 7-1:
The Electromagnetic Spectrum

radio, television, ham radio, walkie-talkies, CB (citizen's band) radio, cell phones, and many other wireless devices.

To "see" something means that some frequency of light hits the object we see, then bounces off that object into our eyes. We can extend the range of what we can see by using electronic devices that respond to frequencies our eyes don't see, and then convert these electronic responses to our visual range. This is most familiar to us in the form of "night-vision" glasses, binoculars, or telescopes, which either take the very faint amount of light available at night (from stars, the moon, or terrestrial sources) and greatly amplify it, or actually take the heat from objects in the form of infrared light, and change it to the frequencies we can see.

Radar and the British in World War II

One of the most significant roles ever played by radar came in the very early days of its development. For several years, citizens of Britain had seen the buildup of Hitler and his armies, and it soon became apparent that his intention was to conquer at least all of Europe. After the fall of France, it was also obvious that it would be only a matter of time before Hitler attempted to conquer England.

For several years prior to the fall of France, scientists, engineers and technicians in England had been working long and hard on radar, as it had the potential to be their biggest ally in the battle against Hitler's forces. If the Royal Air Force had to rely on spotters (citizens with binoculars constantly watching for incoming aircraft), they would have *at best* only a 1- or 2-minute warning, which was simply not enough time to defend themselves adequately, and they knew it. Early tests with radar showed it had the potential to allow them to see far enough to provide a 10- or even 15-minute warning, which was a *huge* difference. This could allow them time to scramble fighter aircraft from multiple bases and even have them *above* the incoming aircraft, which was an ideal attack posture.

Hitler knew something about the development of radar, but his team of developers was a few years behind. This development difference, along with a few other important factors, was key in what came to be known as the Battle of Britain. Repeatedly, Hitler's *Luftwaffe* flew across the English Channel to bomb British air bases, and although the *Luftwaffe* did inflict much damage and were successful in destroying many British aircraft, they were always met by a very significant opposition force and always lost many bombers. Although Hitler's *Luftwaffe* greatly outnumbered the relatively small Royal Air Force (the United States had not yet entered the war at this time), the attrition of German bombers was very costly, and Hitler eventually changed his tactics, and gave up planning an invasion of Britain.

With the *Luftwaffe*, Hitler did continue to inflict damage on many English cities, most notably London, but he never managed to reduce the British defenses to the point that he deemed it safe to attempt an invasion of occupying forces. And there never came a time when a *Luftwaffe* flight was not met by British fighter planes, already airborne and in position to take down as many German bombers as possible. There is no doubt that a very large part of the success of Britain against the greatly outnumbering forces was due to their ability to use radar to see, far in advance, the arrival of an attacking *Luftwaffe* force.

Faint Echoes

Not long after people began experimenting with radio waves, they noticed that in some cases the pulses of radio energy (low frequency light) they sent out were detected twice at the receiver: once when the energy reached the receiving antenna, and again when the energy bounced off a nearby mountain or building and then reached the antenna. The second pulse of energy was noticeably weaker, and it was delayed in time because it took a longer path to reach the antenna.

These echoes of the direct signal were very noticeable in the early decades of television, when stations broadcast over the air and homes received the signals through roof-top antennas. This was known as *ghosting*, because the echoed signal was weaker than the direct signal, and resulted in a picture which featured a horizontally-shifted faint image (ghost) of the direct signal.

Since the concept of radio wave echoes was familiar, engineers thought of ways to put this to good use. One of the earliest applications was in RAdio Detection And Ranging, or RADAR (now just radar).

The amount of time it takes your voice to echo off a nearby building or canyon wall is directly proportional to how far away you are. Sound (at sea level) travels about 770 miles per hour (mph), or 1130 feet per second; if you are 1130 feet away from the canyon wall or building, your echo (if you can hear it!) will take two seconds to return (one second each way). This same principle works with light and radio waves, which travel MUCH faster (about 186,000 miles per second, or about 670 **million** mph). For example, if you pulse a laser at the moon (average distance about 240,000 miles), it will take about 2.6 seconds for the pulse of laser light to reach the moon and return. (By the way, this is one of the experiments set up by Apollo astronauts; they actually left a mirror on the moon so we could do this more effectively!)

If we send a pulse of radio waves out into the empty air, the pulse will not return at all, but continues on into space forever. However, if there is something in the air, a small amount of the energy will reflect off that object and return to the source of the pulse. The amount of time it takes for the pulse reflection to return is directly proportional to how close the object is. If the object is somewhere in Earth's atmosphere, the pulse would return rather quickly, since that is so much closer than the moon. For example, if an airplane is 100 miles away, the echo of the pulse will return in only 5 ten-thousandths of a second (500 microseconds).

These echoes are truly faint. The pulses of energy sent out for radar are intense but brief; they can be from thousands of watts to hundreds of thousands of watts (750,000 W, for example), but they last for only milliseconds to microseconds. The echoes come back in the range of milliWatts to microWatts; and even though these echoes are billions and even trillions of times weaker than the signal sent out, they are strong enough to be detected by very sensitive radar receivers, and from these echoes, an image can be constructed of the object off which they echoed.

Is That All There Is?

Carefully measuring the time from when the pulse is sent out until the echo returns allows radar to know the distance to the object that caused the echo. Modern and inexpensive electronics allows us to measure this time down to a resolution of nanoseconds (billionths of a second), with which we can measure the distance to a resolution of about a foot (which is about how far these pulses travel in 1 nanosecond). Higher resolution is possible with more expensive electronics. So once we detect the echo of the pulse, we can tell quite accurately how far away the object is.

But in the realm of finding things, we care not only *where* they are, but also whether they are moving. In the case of objects in the air, the questions are: Is it moving? How fast is it moving? In which direction? How big is it? What is it? Answering these questions becomes an electronic goldmine, albeit a rather complicated one. The wonder of it is that we can arrive at reasonably confident answers to all of these questions by simply sending out more pulses, then by carefully analyzing the results. We will answer each of these questions in the next sections.

Is It Moving?

The answer to this question is found in something known as the Doppler Effect, named after Christian Andreas Doppler, who in the mid-19th century first proposed an explanation for a commonly observed phenomenon. For example, we notice it any time a moving sound passes by us, such as an ambulance sounding its siren, or a race car at high speed. We always hear the sound *drop* in pitch as the moving object passes us by. This is explained by the fact that sound waves cannot travel faster than their fixed velocity of about 770 mph. So, if a car travels at 100 mph and sounds a siren, the sound waves of the siren are NOT traveling at 770 + 100 mph or 870 mph—they are still traveling at 770 mph—the fastest they can travel in air.

Again, readily observable examples exist to understand this better. In water, the waves travel at a fixed speed. If a small boat is stationary in the water, rocking the boat will cause waves to proceed away from the boat in all directions at the same speed. One can observe these even with a small boat in a bathtub. The distance between the peaks of the waves will be the same in all directions. But move the boat, and it can be seen that the waves at the front of the boat are much closer together than the waves behind the boat. The boat is essentially pushing the waves together.

The same is true of a moving vehicle. If the vehicle emits a sound, the sound waves cannot go faster as the vehicle moves faster, so they get pushed together at the front of the vehicle and spaced apart at the back of the vehicle. Sound waves which are closer together are perceived by the ear as higher in frequency, so as the vehicle approaches, we hear a higher frequency. As the vehicle passes, the frequency drops because the sound waves are farther apart.

Figure 7-2:
A typical radar dish, which can rotate 360° continuously, both sending radar signals and receiving their echoes.

So why all this discussion about moving objects and sound? In radar, what we care about are moving objects which are too far away to be audible! The simple fact is that radio waves (which are of a lower frequency than visible light waves) behave the same way as sound waves, since the velocity of light is fixed, and thus does not change as the velocity of the moving object changes. If we could readily see objects moving at speeds over 10,000 miles per second (36 million mph), we would see their green lights coming at us as blue (shifted *up* in frequency), and going away from us as yellow (shifted *down* in frequency). Since objects moving at this speed are not common on this planet (and that's surely a good thing!), we do not experience seeing such things, but we can often notice this Doppler shift in sound from moving objects.

The application of this principle to radar is similar, but is applied to the *echoes*. If we send a burst of a strong radio signal out into air, it will return in very small amounts from objects that are in the air. If we place our antenna near an airport and sweep (rotate) this antenna continuously, the only echoes we will get will be those bouncing off airplanes in the vicinity. If our sensitivity is such that we can "see" airplanes up to 100 miles away, it works somewhat like this. The strong burst of radio signal is sent out (at the speed of light), and the antenna is then switched to receive mode, to await the echoes. At the end of about 1 ms (1/1000th of a second), the echo from the airplane 100 miles away returns and is detected by the receiver. This information is displayed on the radar screen, and is duly noted by the computer driving the system. As soon as this is done, the next signal is sent out and we again wait for echoes to return. Since it only takes about 1 ms to send out a pulse and wait for the echo to return (from a distance of 100 miles), we can do this about 1,000 times per second. If the sweep takes 5 seconds to go all the way around, that means we can send pulses and receive echoes 5,000 times each rotation.

The echoes from nearby airborne objects return sooner that those from more distant objects. Since the time it takes the echo to return is directly proportional to the distance, and since we know the velocity of the radio waves very accurately, we can determine the distance to the object quite precisely, and can do this many times per second.

If the object is moving *away* from the radar dish, the echo will be at a frequency which is *lower* than what was sent out. Conversely, if the object is moving *toward* the radar dish, the echo will be at a frequency which is *higher* than what was sent out. So by analyzing the frequency of the echo and comparing it to the frequency of the pulse sent out, it can easily be determined if the object is moving.

How Fast Is It Moving?

If an airborne object is not moving (ignoring the fact that this is not very likely!), each echo from that object that is received at our antenna will occur at the same time, and at the same frequency of the original pulse. However, if the object is moving *toward* the antenna, each echo will occur in successively less time, and at a higher frequency, as discussed above. For instance, if we first receive an echo from the airplane at 100 miles distance, the echo will be received 1.075269 ms after the burst was sent. If the plane is flying 200 mph and we send out pulses every 2 milliseconds, the next pulse will give us an echo in 1.074672 milliseconds, which is 597 nanoseconds (billionths of a second) *less* than the time for the first pulse to return. Knowing the time between when the pulses were sent, and knowing the difference in the times when the echoes were received, allows us to determine the speed of the object. Although it seems nearly incomprehensible that we can measure such tiny increments of time, it is by no means difficult using today's technology. In fact, using relatively inexpensive electronic circuits, we can readily measure increments of time as small as 1 picosecond (1 trillionth of a second)! (And we can do even better, if we are willing to spend the money; increments as small as 1 femtosecond (1 quadrillionth of a second) have been very reliably measured!)

As one would expect, the differences in the times for the echoes to return is directly related to the speed of the object. In the example above, the difference in the time for the echoes was 597 nanoseconds for an airplane traveling 200 mph. If the airplane were traveling twice as fast (400 mph), the difference between the two echoes would be 1194 nanoseconds, or twice the original difference.

In Which Direction Is It Moving?

In the examples given above, the airplane was flying *toward* the antenna. How would this example change if the airplane was flying *away* from the antenna? As one might expect, the numbers remain the same, but the relationship reverses! That is, the pulses would arrive in successively *more* time, instead of less time. If our 200 mph airplane is 100 miles away, but traveling *away* from our antenna, the second echo will arrive in 1.07587 ns, 597 ns *later* than the first pulse.

There is admittedly an oversimplification in the above examples, in that we have only considered an airplane traveling either directly *toward* or *away* from our antenna. Thankfully, reality is only a bit more difficult to handle. By using a kind of math known as vector algebra, we can take into account all three axes of movement possible for the airplane. This allows us to determine the velocity and direction of the airplane, no matter which way it is going with respect to our antenna.

How Big Is It?

If one were to shine a very bright flashlight at a distant mirror of very small size, one would only see a very small, dim reflection. But increase the size of the distant mirror, and the size and brightness of the reflection will also increase. This same relationship also holds true for radar. A very small airplane will return only a very faint echo; a much larger airplane will return a much stronger echo. So, the brightness of echo we receive tells us much about the size of the object that caused the echo.

But again, there is an oversimplification in this statement, since it is easy to see in the above example that if the same small mirror were moved twice as far away, the reflection would be smaller still. Thus, *distance* also plays a role in the strength of the echo. But this is another place where we can use mathematics to correct for this problem. Since we know the distance (by measuring the time for the echo to return), we can use this information to tell us about the true size of the object.

What Is It?

This question is clearly one that we would love to be able to answer. Unfortunately, such information is much more difficult to come by. The first approach used to answer this question is still one of the most commonly used, since it is quite effective. If there were an infinite variety of items that could be airborne, there is not much we could do to answer this question. However, since this is clearly not the case, there is much we can do to narrow down the possibilities.

If the return echo tells us the object is quite large, and that it is traveling at 200 mph, we can immediately tell it is not a single-engine airplane. Furthermore, if we know what airplanes we can *expect* to be in our surrounding airspace, we can choose from a much smaller list of possibilities. Once positive identification is made, we can attach this information to the echoes received from the object, allowing the object and its associated information to remain together until the object passes out of our airspace.

The idea of high-resolution radar holds much promise, but is much more difficult. It is true that different parts of an airplane will produce different amounts of echo, and that these different parts of the echo will arrive at the antenna at slightly different times. Turning this information into a 2-dimensional, or even a 3-dimensional, image of the object requires MUCH greater precision and resolution in the radar system. While such systems have been demonstrated, they are not yet fully practical. However, where such systems can be justified (as in the life- and mission-critical applications in the military), high-resolution radar systems have been deployed, and have been shown to provide very valuable additional information.

A Word About Sonar

Sonar (originally an acronym for SOund Navigation And Ranging) operates on the same basic principles as radar, except that it uses sound pulses instead of electromagnetic pulses. Seawater is quite conductive, and electromagnetic waves do not travel well in it, so radar is almost completely ineffective. But sound travels extremely well in water, and at a much higher velocity (about 3315 mph, as opposed to about 770 mph in air), and so sonar has been used for many decades as a very effective way to "see" through water, for ships, submarines, and airplanes seeking to find other ships and submarines. Using the same principles as radar, these sound pulses can determine the position, velocity, direction, and approximate size of other underwater objects.

What About Stealth Technologies?

Stealth technologies are not part of the marvel of electronics, but some mention should be made here of their significance. Stealth technologies deal entirely with materials and physical design and construction techniques that absorb or highly scatter radar pulses, so that the amount of signal that actually echoes to the radar dish is much smaller than it would otherwise be. Note that the reflection is much *smaller*—it is not completely eliminated. But such a small reflection can easily be mistaken for something it is not, or it can be missed entirely, and therein lies its value.

Today's stealth fighters and bombers have "radar cross-sections" (a term basically meaning the radar-equivalent size of the object) equal to an object the size of a golf ball or a soccer ball, respectively. While this is widely known in military organizations, stealth technology is still a major advantage because that makes them so much harder to detect. This same technology is also being applied to ships, to reduce both their radar AND sonar cross-sections.

Chapter Take-Aways

The next time you look overhead and see an airplane passing by, think about the fact that if your eyes were as sensitive as radar systems, you would have seen that airplane long before. And you would have known its velocity, direction, and approximate size. That would certainly make your vision something like that of Superman's.

Radar is also at the heart of the air-traffic control systems for commercial airplanes around the world. Each time you take off, you can thank radar for allowing the air-traffic controllers to know when it is safe for your flight to take off. And the same is true each time you approach an airport for landing—without radar, it would be more than a mess—it would be so unsafe that people would certainly avoid it!

One other thing worth mentioning about radar, and which will be covered in more detail in Chapter 16 (Automobiles)—radar is already finding a very significant role in automobiles, and this role appears to be growing as technology advances and continues to come down in cost. Stand by for the car of the future!

CHAPTER 8
FROM THE NETHERMOST PARTS:
DEEP SPACE PROBES

Just How Far is Far?

At the present speeds mankind can attain for space travel (a mere 70,000 mph), interplanetary exploration is not something we will do with humans, with the possible exception of Mars. Even this relatively short trip is planned to take over 1½ years (each direction!) Given this limitation, it is not practical to send humans out to explore beyond Mars, and so we send space probes such as Voyager 1 and 2. Since these two probes epitomize our success in interplanetary exploration, they will be used as our example throughout this chapter.

These hardy little probes were launched in 1977, and were originally scheduled to rendezvous with only Jupiter and Saturn. Due to a

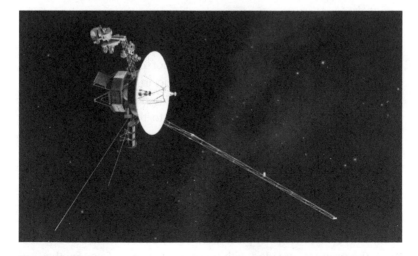

Figure 8-1:
The Voyager 2 space probe—truly one of our most amazing communication accomplishments.

unique combination of events, Voyager 2 was able to be sent farther on to Uranus and Neptune, an opportunity that would not repeat for many decades. Fortunately, this intrepid explorer was up to the task. When Voyager 2, after 15 years of traveling, finally reached Neptune, it was approximately 2 billion miles from Earth! This was so far away that radio signals used to communicate with it, even at the speed of light, took almost three *hours* to cover the distance!

A Bit of Background

A deep-space probe is different from a satellite in some very fundamental ways. First, it must be powered by something other than solar panels. Since the effective brightness of the sun falls off with distance, solar panels don't have enough sunlight to be effective beyond Mars. So what can power one of these little probes for more than 15 years? Not even our best, most expensive batteries are capable of providing the needed energy for so long. What about fuel cells? These almost magical devices are capable of producing electricity from hydrogen and oxygen, with pure water as the byproduct. While this is greatly useful for manned missions such as shuttle trips, it is impossible to provide enough hydrogen and oxygen to last for so many years. In fact, there is only one known source of energy compact and powerful enough to meet this need: nuclear energy.

Plutonium is indeed a rare element. The 94th element on the periodic table of elements, plutonium does not occur naturally, and is an unstable element. It can only be made in nuclear reactors and particle accelerators, and once made, it immediately begins to decay. As it decays, it gives off energy according to the most famous of all Einstein's equations, $e = mc^2$. This equation says that the energy given off is equal to the mass that is converted to energy, multiplied by the *square* of c, which is the speed of light. Since the speed of light is a rather large number, the speed of light *squared* is *huge!*. Basically, this means that a very small amount of matter, when converted entirely into energy, provides a great deal of energy!

So, with a small cargo of plutonium (only about the size of a grapefruit), the probe is provided with a constant source of energy that can last for decades. A device known as a thermopile directly converts this energy, in the form of heat, into electricity. For the Voyager probes, this nuclear energy source was capable of approximately 300 watts of continuous power for over 40 years!

Another way in which deep space probes are different than satellites is their *non*-serviceability. While it is extraordinarily expensive and often impractical, it is nevertheless *possible* to service satellites in space, as has been demonstrated with the space shuttle. However, this is not an option with deep space probes. If there is a design flaw, or if there is a premature failure of some kind, we must simply live with the consequences. Accordingly, there is extra care taken to make sure that these probes do not have problems, and to allow them to be adjusted from Earth.

As is the case with most things today, Voyager 1 and 2 are controlled by an on-board computer. Computers are very versatile, and the fact that they can be reprogrammed is one of their greatest strengths. Deep space probes are designed to be reprogrammed from Earth, which has been essential to their success. While it is true that most, if not all, satellites can also be reprogrammed from Earth, the degree to which this type of design permeates the deep space probe is greater than it is for satellites.

The last way in which deep space probes are different from satellites[1] lies in their singular mission focus, due to the very high costs of launching them. Satellites orbit the earth in many different orbits; the higher the satellite's orbit, the more it costs, per pound, to launch the satellite. Deep space probes do not orbit the Earth at all, so the effective height of their orbit is infinite. This means they are MUCH more expensive, per pound, to launch, so every pound of cargo must be thoroughly justified. Only that which is absolutely essential to the mission can be included; redundant hardware is minimized as much as possible.

What Does It Take to Communicate Over Billions of Miles?

Essentially every tool known to electronic engineers today was employed in some form to allow Voyagers 1 and 2 to communicate over a distance of two billion miles. Some of the most significant tools,

---

[1] There are admittedly other ways in which deep space probes differ from satellites, but these three differences will suffice for this discussion.

still employed today, include high-gain antennas, common-mode rejection, digital modulation, and error-correction coding. Without these, communication over these vast distances would simply have been impossible. Even with these tools, it seems nothing shy of a miracle that a signal so weak could travel so far and convey such accurate information. These rather remarkable technologies will be explained in this chapter.

Just How Weak IS That Signal?

Early radio transmitters used extremely powerful transmitted signals (up to and even exceeding 100,000 Watts!), but even then their signals could only be detected for a few hundred miles. Over the years, some improvements were made with antennas, but the major improvements were made with the receivers, as they became more and more sensitive to the weak signals that arrived. How weak are these signals?

A typical AM radio station in the USA will transmit at 50,000 Watts, yet only a few dozen miles away, a typical AM radio will pick up a signal of only about 50 microWatts, which is 50 millionths of a Watt. That's not even enough energy to run the power-indicator LED of your computer! Yet as small as they are, those 50 microWatts are HUGE in comparison to the signals received from deep-space probes. Voyager 1 and 2 had a transmitter of only 20 Watts—over 2,000 times less power than a typical AM radio station. And the distance their signals had to travel—billions of miles—is tens of millions of times greater than the distance from a radio transmitting tower to a receiving radio[2]. So what do we get here on Earth? Only a few attoWatts! (that's 10^{-18}!) To put this into perspective, we could liken the signal from a deep-space probe to the light from a large, 4-battery flashlight, and the detectors on Earth to an eye that can see that flashlight being turned on and off at a rate of thousands of times per second, at a distance of billions of miles. Although it is hard to conceive of a receiver sensitive enough to

[2]The NASA website informs us that Voyagers 1 and 2 are STILL communicating with Earth, more than 30 years after their launch, and that they should be able to continue this until 2020. They are presently more than 10 billion miles from the Sun, and radio signals from them take over 15 *hours* to reach Earth—at the speed of light!

actually detect such an incredibly weak signal, such systems exist, and function amazingly well.

The next sections will address some of the major electronic advances that have made this amazing feat possible.

High-Gain Antennas

As most of us are familiar, an antenna is a device used to receive and transmit electromagnetic signals in the form of radio signals. In most applications, an antenna is intended to receive and transmit signals in all directions. This is termed *isotropic*, which means the same in all directions. If the antenna of a cell phone were not isotropic, you'd have to be facing the cell tower for it to work, and that's not very convenient! However, this convenience comes at a cost, which is the efficiency of the communication link. In order to be isotropic, the cell phone antenna must transmit signal in all directions. This means that the majority of the signal from the cell phone antenna does *not* reach the desired cell phone tower, and instead goes off in all other directions, bouncing off buildings, mountains, and heading into outer space. But since there is enough of the signal going in the direction we want, we are okay.

There are other applications in communication for which an isotropic antenna is the *opposite* of what you want. If you are trying to communicate from a deep-space probe to Earth, you can't afford to waste so much of your signal. Since you know in what direction you should send the signal, the antenna should be of the type that does this as effectively as possible. For the range of frequencies used on deep-space probes such as Voyagers 1 and 2, this type of antenna is a high-gain antenna in the shape of a familiar satellite dish.

The directionality or gain of a dish antenna is proportional to its size. Accordingly, it would be very desirable to have a HUGE dish antenna on our deep-space probe. But again, launch costs and size constraints limit what we can do, so we must be satisfied with a simple 2-meter dish antenna, giving a gain of about 20 dB, or providing 100 times more signal in the direction we want, compared to an isotropic antenna.

There is one other thing that is essential to the functioning of a high-gain antenna, and that is very precise directional orientation. Basically, the space probe must be able to know *exactly* where to point the antenna; a fraction of a degree off will mean complete loss of communication! Accordingly, the probe is equipped with highly accurate navigational equipment as well as very precise attitudinal thrusters, allowing it to always keep its antenna pointed the right way. Without accurately-pointed, high-gain antennas, communication with deep-space probes would not be possible.

Common-Mode Rejection

In the author's opinion, this is a classic case of taking a lemon and making lemonade from it. Noise is inherent in all electronic systems, and a large measure of the degree of success of an electronic system lies in its ability to deal effectively with this noise. Examples of noise in older analog systems include the tape hiss common to 8-track and cassette tape systems, the scratches, ticks, and pops common to phonograph records, and the humm or buzz common to public address systems.

If you were to build a dish antenna and point it skyward, then listen to and watch the resulting signal on your television, you would notice a lot of random noise, sometimes referred to as "snow" in a television signal, or white noise in an audio signal. This noise coming from outer space is the noise that is competing with the desired signal from our Voyagers 1 and 2 deep-space probes. Where does this noise come from? It comes from stars, nebulas, pulsars, quasars, and other types of astronomical objects. But how is it possible for noise to travel through space?

The kind of noise we are dealing with here is *electromagnetic interference*, or EMI. It is a conglomeration of many different frequencies of *electromagnetic radiation* (EMR), which is not to be confused with nuclear radiation—they are different in almost every way. EMR is nothing but waves of electrical and magnetic energy traveling through space, and we know this form of radiation by many names, depending on its frequency (or wavelength)—see Figure 7-1. Looking up at the Sun, we call this EMR *light*, or visible light. The various bands of wavelengths in visible light are readily shown to us each time we see a rainbow; the lowest visible frequency band is red (always at the outer edge of the rainbow), and the highest visible frequency band is violet (always at the inside edge of the rainbow). Frequencies below the red band are known as *infrared* (infra means below); frequencies above the violet band are known as *ultraviolet* (ultra means above or beyond).

Frequencies below and above these include microwaves and radio waves, then x-rays, gamma rays, and cosmic rays, again as depicted in Figure 7-1.

Fortunately, our antenna and radio receiver back on Earth are not sensitive to ALL these wavelengths—that would be a bit overwhelming! But they ARE sensitive to a large part of the radio waves and microwaves, and thus the EMI mentioned in the previous paragraph, coming from many astronomical sources, threatens to overwhelm the meager signal from our little deep-space probes. In fact, this astronomical noise is many orders of magnitude *larger* than the signal we desire. To hear a signal so faint amidst noise so large is truly an amazing feat, and could be likened to standing near the speakers of a rock band at a concert, while listening to someone over a hundred feet away whisper to you, and being able to hear them above the rock band. And this is not even exaggerating.

So how is this magic feat accomplished? Through a technique known as *common-mode rejection*, which in its basic nature is really quite easy to understand. It uses a special type of amplifier known as a *differential* amplifier (see Figure 8-2), which amplifies the *difference* between the two signals at the input (left side) of the amplifier. But one of the inputs to this differential amplifier is unique, in that it inverts the signal coming into it. The end result is such that if you put a signal on the + input, and the exact inverse of that same signal on the − input, you will get no signal at all on the output.

Okay, so that was easy to understand, but how does it help us detect our faint signal amidst all the noise? The answer is to take the

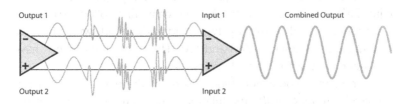

Figure 8-2:
A differential amplifier amplifies the inverse of the signal on the − input, and the + input, giving a single signal which combines these differential signals.

noise signal and put the exact inverse of it on the − input of the differential amplifier (inverting an electronic signal is very easy to do), then take the same noise signal, plus the desired deep-space probe signal, on the + input of the differential amplifier. The result: the differential amplifier amplifies the difference between the inverse of the − signal and the + signal. The inverse of the − signal is the noise; the + signal is the noise plus our desired signal, so the difference (see Figure 8-2 again) is our desired signal! Only the desired signal is amplified—the noise isn't amplified at all! This means that the noise is actually *decreased* in relative amplitude by the same amount that the amplifier *increased* the amplitude of the desired signal! This quite remarkable electronic feat allows us to accurately detect a desired signal which is trillions of times smaller than the surrounding noise—this is truly amazing!

Digital Modulation

As covered in Chapter 6, without modulation (changing), it is not possible to send information over an electronic carrier signal. This is readily perceived if one considers how information is carried over sound waves. If someone were to say only the same one sound all the time (not modulating), no information could be transmitted by sound. It is only by changing (modulating) the sounds into meaningful chunks (words, then sentences) that information can be conveyed.

The same is true for sending information over electromagnetic waves. If we have a carrier, we must modulate that carrier in some way for information to be conveyed. Two of the oldest methods of modulation involve modulating its amplitude (known as *amplitude modulation*, or just AM) or modulating its frequency (known as *frequency modulation*, or just FM). AM is simpler than FM, so it came first, followed a couple of decades later by FM; thus the two major radio bands with which we are familiar are known as AM and FM. These two radio bands have since been with us for many decades.

Together, the modulation types used by these two radio bands can also be termed *analog* modulation, which simply means that the modulation is accomplished in the same way that our voice produces sound—in a continuous variation, changing continuously from one amplitude or frequency to another.

106

Digital modulation could be likened to singing, rather than simply talking. In talking, we can use any range of frequencies we wish to communicate our message, though we usually stay within a fairly small range. But in singing, we are restricted (in the tradition of Western music) to 12 frequencies: A through G, with 5 half-steps known as sharps or flats, depending on the key. If a singer does not hit the frequency (*pitch*) just right, we hear it as being off-key, and it sounds unpleasant. Thus, singers are restricted to 12 frequencies, and their integer multiples.

If a message to be communicated is restricted to few values, it is much easier at the receiving end to determine what values were actually sent by the transmitter. For example, if two people who are both hard of hearing were to restrict their conversations to 3 or 4 words, it would be rather easy for them to learn to decipher which word was spoken each time they heard something. The biggest challenge for any listener in a noisy environment comes when they have no idea what word, of the hundreds of thousands of possibilities, is being spoken. This is where digital modulation comes through with flying colors.

In the simplest form of digital modulation, we are restricted to only *two* possible values (amplitudes, frequencies, or phases), and they are different enough that it is quite easy to tell them apart. For instance, if a person were showing a different card every second and you were tasked to determine what color card the sender was showing each second, your task would be quite simple if the sender only had two choices: black or white. It is this restriction to only two very different values that gives digital modulation its great ability to communicate information in a very noisy environment.

Error Detection and Correction

It is hard to pick a favorite among the amazing technologies that have made deep-space communication possible, but if I were pressed to do so, I would choose this one. Its power is only understood after a few examples, and after starting with the simpler forms of error detection.

The value of it can be readily perceived. Wouldn't it be great if the receiver of a signal could know right away if they had misperceived a word or symbol of the message? Then they could tell the sender there was a problem, and something could be done to fix the problem. But how can this be done? First, by taking advantage of the fact that we're dealing with digital modulation, so we know there are only specific values that could be sent. And second, by sending a little extra (redundant) information about the message. To clarify, consider the example given in

Data				Sum
4	2	8	9	23
1	5	3	7	16
9	0	4	6	19
8	6	9	2	25

Figure 8-3:
Data with simple sums for each row.

Figure 8-3. In this example, we are restricted to sending single digits, with values from 0 to 9. Every time we send 4 of these digits, we follow that 4th digit with a pair of digits, representing the *sum* of the previous 4 digits. The transmitter is easily able to calculate this extra information about the message, and the receiver can readily check the digits as they come across. If one of the sums doesn't agree with the received sum of the previous 4 digits, the receiver knows an error has occurred, and corrective action can be taken.

PARITY

The preceding form of error detection is done digitally using a concept known as parity, which is actually just a binary form of adding.

Data								Parity	
								Even	Odd
1	0	1	1	0	1	1	0	1	0
0	0	1	0	1	1	0	0	1	0
1	1	1	0	0	1	1	1	0	1
0	0	1	0	0	0	1	0	0	1

Figure 8-4:
Digital data with simple parity "sums" for each row.

This is shown in Figure 8-4. In these rows, the far right two columns represent the two types of parity for the previous 8 bits (a bit is either a 0 or a 1). For even parity, the total number of ones in each row, *including* the parity bit, must be *even*. For odd parity, the total number of ones in each row, *including* the parity bit, must be *odd*. So, for the first row, which contains five ones, the parity bit must be 1 for even parity, giving the total row an even number of ones (six). The second row has three ones, so the parity bit must again be 1, giving the total row an even number of ones (four). And in row 3, which has six ones, the even parity bit must be 0, giving the total row six ones, again an even number.

Odd parity can also be used, and this is also shown in Figure 8-4. There is no inherent advantage to either even or odd parity, but once the decision is made which to use, it is no longer arbitrary for the receiver to choose even or odd parity—the receiver must choose the same parity as the transmitter.

Looking at Figure 8-4, it is easy to see what would happen if any of the bits were to be mistakenly reversed by the receiver. The parity for that row would be incorrect, and the receiver would know an error had occurred. This is also true if one of the parity bits is mistakenly reversed at the receiver. However, there is a hole in this approach, and it is a rather LARGE hole. If *two* bits in any row were to be mistakenly reversed in any given row, the parity would not indicate the presence of an error! As shown in Figure 8-5, this is also true for any even

Data								Tx Parity Even	Odd	Rx Parity Even	Odd
1	0	1	1	‡	1	1	0	1	0	0	1
0	0	‡	⊖	1	1	0	0	1	0	1	0
1	⊖	1	0	‡	⊖	1	1	0	1	1	0
‡	0	1	‡	0	0	⊖	‡	0	1	0	1

Figure 8-5:
Rows of bits, their parity, and cases where multiple bits are in error in a single row. In row 1, the single-bit error is detected by the parity. The three bit-errors are detected in row 3. However, the 2 bit-errors and the 4 bit-errors in the second and fourth rows would not be detected by the parity.

number of bit mistakes—they cannot be detected! So, while parity is easy to implement and detect, it is not as powerful as we would like it to be—it misses too many of the cases where multiple bits have been mistakenly reversed.

One disadvantage of parity is the additional (redundant) data that must be transmitted—the parity bits themselves. For the example in Figure 8-4, there is one parity bit for every 8 bits, which means 1/9 of the data is redundant—11.1%. This 11.1% is called *overhead*.

CRC

A second approach to parity, with significantly more detection power, but also with more complexity, is known as CRC, or Cyclic Redundancy Check. As with the parity error detection described in the previous paragraph, it also uses binary addition to generate information about the data. But for CRC, one of the big advantages is that it is able to detect errors involving multiple bits—in fact, any number of bits in error—within a certain probability, and the probability that an error would be missed is quite small for 16-bit parity (about 15 parts per million, or ppm). And the overhead is also quite small—only 16 extra bits need be added, almost irrespective of the size of data file. CRC creates what is known as a checksum, which is computed using iterative feedback—each new bit causes a different checksum to be generated. CRC is presently very widely used for error detection in digital communication. An example of a 16-bit CRC generator is given in Figure 8-6.

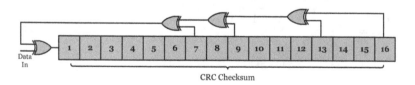

CRC Checksum

Figure 8-6:
An example of a circuit for calculating a 16-bit CRC checksum. By today's standards, this circuit is extremely simple, involving only 4 exclusive-OR gates and 16 flip-flops (storage elements).

For 16-bit CRC, the number of redundant bits is always 16, but CRC can be computed over short blocks of data or large blocks. If it is calculated over 100 bits, there would be 16 redundant bits in 116, or 13.7% overhead, which is not great. But if a 16-bit CRC checksum is calculated over 65,536 bits (4 kbytes), then the overhead drops to only 0.0244%—and that is great! (because of such a very low overhead).

But in the end, as good as CRC checksums are, they can only tell that an error has occurred—they cannot fix the problem. That capability is reserved for the last type of error detection.

FEC

The final type of error detection, known both as Forward Error Correction (FEC) and as Error Correction Coding (ECC), is much more complicated than the previous two types, and it involves more overhead and much more computation. But its power is not to be underestimated—it is amazingly powerful, and has been adapted for use in nearly all forms of digital communication and data storage. Its power comes from the fact that it goes one major step beyond the previous two types of error detection—it actually can tell WHICH bit (or bits) were in error. And since we're dealing with a binary system, as soon as we know WHICH bit was in error, we also know how to fix it—just change it to the opposite value!

The math behind practical forms of FEC in use today is fairly complex, but the concept is readily grasped with an example, as shown in Figures 8-7a & b. Each row has simple parity added (as described in the **Parity** section previously). Likewise, each column also has parity added. While this results in more overhead, the advantage it gives us is that we can identify WHICH bit was in error in the block of data. Knowing this allows us to fix the offending bit.

As before, having multiple bits in error in the same row (or column) causes rather severe problems for FEC—it prevents us from being able to determine which of several bits were actually in error. This could be fixed with additional parity bits, and this is one of the solutions that is often used. However, even multiple additional parity bits cannot help us identify all the bad bits if the bad bits are too close together. And the nature of errors in digital data is that they are almost ALWAYS close together—in *bursts*. Whatever the event was that caused one bit to be wrong is also very likely to cause several adjacent bits (in the same burst) to be wrong. This bursty nature of digital errors is readily solved by another brilliant yet simple solution: interleaving.

An example of interleaving is given in Figure 8-8. We simply change the order of the bits before we send them out, so that bits that were originally adjacent are not sent adjacent to each other. Thus, when a burst error occurs, it does NOT affect originally adjacent bits. In the example of Figure 8-8, a burst error 8 bits in length would wipe out bits in 8 different bytes of the 64-bit stream, but none of these bytes would have more than 1 bit in error, which is easy to detect and

Data								Parity
1	0	1	1	0	1	1	0	1
0	0	1	0	1	1	0	0	1
1	1	1	0	0	1	1	1	0
0	0	1	0	0	0	1	0	0
1	1	0	1	1	1	1	0	0
0	0	0	1	0	0	1	0	0
1	1	1	1	0	1	1	1	1
1	0	0	1	0	1	1	0	0
1	1	1	1	0	1	1	0	0/1

Data								Parity
1	0	1	1	0	1	1	0	1
1	0	1	0	1	1	0	0	0
1	1	1	0	0	1	1	1	0
0	0	1	0	0	0	1	0	0
1	1	0	1	0	1	1	0	1
0	0	0	1	0	0	1	0	0
1	1	1	1	0	1	1	1	1
1	0	0	1	0	1	1	0	0
0	1	1	1	1	1	1	0	0/1

Figure 8-7a:
An example of block coding for error detection and correction.

Figure 8-7b:
An example of the same block of data and parity, with two bits in error. It is possible to identify exactly WHICH bits are in error.

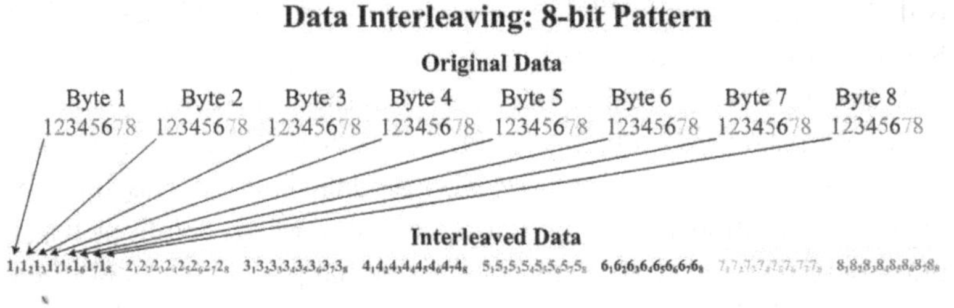

Data Interleaving: 8-bit Pattern

Original Data

Byte 1	Byte 2	Byte 3	Byte 4	Byte 5	Byte 6	Byte 7	Byte 8
12345678	12345678	12345678	12345678	12345678	12345678	12345678	12345678

Interleaved Data

$1_11_21_31_41_51_61_71_8$ $2_12_22_32_42_52_62_72_8$ $3_13_23_33_43_53_63_73_8$ $4_14_24_34_44_54_64_74_8$ $5_15_25_35_45_55_65_75_8$ $6_16_26_36_46_56_66_76_8$ $7_17_27_37_47_57_67_77_8$ $8_18_28_38_48_58_68_78_8$

Figure 8-8:
An example of 8-bit interleaving. Real systems will often interleave many thousands of bits. Such circuitry is relatively simple.

correct. Fortunately, the hardware required to implement interleaving is very simple and efficient, and this innovation allows us to use less complicated FEC codes, and requires less overhead.

The power of FEC is best grasped with an example. FEC of a type known as Reed-Solomon is widely used in optical disc storage. In the case of DVDs, it requires approximately 25% overhead (meaning that of every 100 bits, approximately 25 of them are FEC bits). In reading back the data from a DVD, an error typically occurs once in every 200 bits. This may not sound bad, but keeping in mind that the data rate we're dealing with here is 11 Mbps (11 million bits per second), this then means we would typically experience 55,000 errors *every second!* Clearly, this is an unacceptable error rate. But with the 25% FEC bits added in, the playback circuits can actually *detect and fix* these errors, improving the actual error rate by approximately *18 orders of magnitude!* Thus, instead of experiencing an average of 55,000 errors each second, we are able to reduce this to only experiencing an average of one error every *50,000 years!*

As Applied to Deep Space Probes

Many great tools have been discussed in this chapter, without which deep space communication would be impossible. But even with high-gain antennas, common-mode rejection, and digital modulation, deep space communication would not be possible without FEC. The number of errors that occur in the transmission, without FEC, would overwhelm the system and no meaningful data could be received. Accurate sources of the exact numbers are very hard to find, but apocryphal sources give some insight. It is thought by some that there is almost 90% redundancy—of every 100 bits, 90% are redundant FEC bits. But as with DVDs, these redundant bits dramatically alter the final number of errors to the point that amazing, digitally perfect pictures were received of Neptune and many of its moons, even at the incredible distance of about 3 billion miles!

Chapter Take-Aways

It is unlikely that the reader will have been involved intimately in any of the deep space probes that have initiated from our planet, since the percentage of such people is surely quite small. So what other more familiar applications are there for all the advances developed for this rather arcane application? The answer: several!

If it weren't for high-gain antennas, microwave communication links (used for phone communication, some video, and lots of Internet communication) would not be possible, nor any satellite communication—and there are **thousands** of satellites, and they are very heavily used.

If it weren't for common-mode rejection, digital modulation, and error detection and correction, none of today's digital communication would be possible. This includes digital radio, HDTV (digital television), high-speed optical links (which are the backbone of the Internet), cell phones, cordless home phones, and Internet links. Additionally, data storage devices such as flash memory, DRAM, SRAM, hard-disk drives, optical disc drives, and storage for digital cameras and MP3 players—all these would be impossible!

So the next time you use any of these digital devices, just think of the amazing complexity that allows the errors in your data to be fixed, on the fly, without any effort on your part! The amazing advances developed for deep-space probes such as Voyager 1 and Voyager 2 have truly proliferated and blessed our lives in many ways. So many of the beneficial communication devices and data storage devices that we take for granted today are enabled by these technologies. In developed societies, rarely a day goes by that we do not use many of these devices several times!

CHAPTER 9
THE EPITOME OF VERSATILITY:
THE COMPUTER

The Need for a Computer

One of the earliest recognitions of the need for something which could produce error-free tables of calculations came from the field of navigation. Prior to the development of the computer or GPS (see Chapter 15), navigators were dependent on large tables of calculations, usually printed out or laboriously copied by hand. But whether printed or copied by hand, such large tables were routinely plagued by typesetting or copying errors, and such errors were almost impossible to detect. How do you find an error of a single digit in a table of numbers? It was a terribly difficult problem to solve, and errors sometimes resulted in shipwrecks, with associated loss of life and cargo.

This dependency on navigation tables, coupled with scattered yet almost undetectable errors, led many early inventors to spend time trying to find a way to mechanically produce these tables without human intervention. Some of these efforts are well described in books such as "The Man Who Knew Too Much: Alan Turing and the Invention of the Computer" (by David Leavitt) and "Computers: Mechanical Minds" (by Don Nardo).

Another early area where computers were considered an essential development involved the aiming of artillery. Newton's laws of motion and the associated equations had given us the majority of what we needed in order to accurately predict where an artillery shell would land, but solving these equations was not trivial, and solving them in time to actually use the results in battle was impractical. Often tables were used, as in navigation, but with the same problems: such tables were inevitably riddled with almost undetectable errors.

More mundane were the needs of those in business who were constantly having to calculate balances, taxes, interest, net worth, and the myriad other sums and differences required to keep track of a business's books.

Early Developments

Early developments in the field of computing included the abacus; a calculating clock invented by Wilhelm Schickard; an adding-only calculator invented by Blaise Pascal and known as the *Pascaline*; a 4-function calculator invented by Gottfried Wilhelm Leibniz and known as the *stepped recokoner*; a power loom whose patterns could be programmed on punched cards, invented by Joseph Marie Jacquard and referred to as the *Jacquard loom*; a calculating machine for computing navigation tables invented by Charles Babbage and known as the *Difference Engine* (which was never finished); a programmable calculating machine also invented by Charles Babbage, known as the *Analytic Engine* (and also never finished); a card reader for automating the 1890 U.S. census, invented by Herman Hollerith, and known as a *Hollerith desk*.

Each of the devices mentioned above added significantly to our ability to compute (usually meaning add, subtract, multiply and divide) in an automated, error-free way. As time and technology advanced, the idea of making a device which was truly programmable became the most important. It became apparent that making a device which could simply perform the four basic functions of math would not satisfy all needs, and that the programmability of such a device would greatly enhance its capabilities. It is therefore very sensible that the first true computer of our era was a *programmable* computer, which enabled it to be used for any application its users desired.

PROGRAMMABILITY

This begs the question why programmability was so desirable. Probably the best answer to this question may be found in the venerable Swiss Army Knife. Although it is not the slimmest knife to be found on the market, it is renowned for being the most versatile. By closing one of its appendages and opening another, it becomes a different tool—essentially being reprogrammed just by changing the open appendage.

So it is with a computer. If a computer is purpose-built to solve a given set of problems, it becomes a single tool in the hands of the computer operators. It is very useful for that application, but it has no other applications. But if it is a programmable computer, it can be reprogrammed to solve any kind of problem, to be used for any type of application, simply by loading a new program for the computer to execute. This versatility makes it much more valuable, and therefore more useful to a wider audience.

RESPONDING TO EXTERNAL STIMULI

As computers began to be developed, one other attribute quickly rose to the top of the desirable characteristics for a computer: the ability to respond to external stimuli. It's great for a computer to be able to produce mathematical solutions, but what if a computer could also change the solution as a function of an important variable, such as temperature or wind velocity? What if the computer could wait for external input before proceeding to certain solutions? This would greatly increase the versatility of the computer. And because this was rather simple to add to the attributes of a computer, it was an early improvement. How a computer responds to external stimuli will be described in a following section; first let's understand how a computer can be so incredibly versatile.

The Inside of a Computer

In its simplest terms, a computer is an electronic instruction executor. Figure 9-1 shows the basic organization of a computer, with the CPU (Central Processing Unit) being the heart of the show. The CPU can put information in memory or get information from memory; it can also receive input from the input devices, and provide output to the output devices.

The input and output devices of a computer are among the most familiar and easily understood parts. Most readers will have used a keyboard and a mouse extensively, and they are very familiar with how the computer responds to them. The same is true for the output devices, the most important of which is the monitor. Working with a computer nearly always involves giving it input through the input devices, and seeing what it does in response through the output devices.

The memory of a computer is a little less obvious, as many readers may never have actually seen what it looks like. In reality, it is

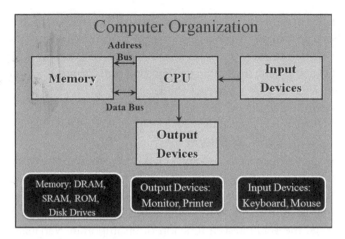

Figure 9-1:
The basic organization of a computer.

somewhat anticlimactic to see the memory of a computer, as it is impossible to see anything happening in it, even when it is installed. Figure 9-2 is a picture of a single in-line memory module (SIMM), which contains the main memory, or DRAM (dynamic random-access memory), of a computer. But if the DRAM of a computer stores the information needed by the CPU, why does a computer also need a hard disk drive (HDD) or optical disk drive (or both)?

The answer to this question brings up one of the most common realities in all of engineering: tradeoffs. It is extremely rare that any product

Figure 9-2:
An example of the main memory in a computer.

designed to meet a given set of requirements will meet each requirement in the optimal way. This is why, for instance, there are so many kinds of cell phones with so many different feature sets in each. None is an optimal solution in all aspects, and each involves tradeoffs. For example, if you want a smaller, lighter cell phone, you may have to sacrifice battery life, since the time a battery lasts is a direct function of its size.

Figure 9-3 shows a diagram of the classic memory tradeoff of access time versus storage capacity. All computer users would love for their computer to be blazing fast, and to be able to store everything they ever wanted to store. The speed of a computer is inversely related to the *access time* of the computer's memory. Access time is simply the time between when the CPU requests some data from memory, and when the CPU receives that data from memory. If the access time is short, then the CPU doesn't have to wait very long, which makes the computer faster.

The storage capacity is simply how much data the computer can hold. There are two main data storage types with which most computer users are familiar: the main memory (the DRAM), and the disc drives (hard-disk drives or HDDs, optical disc drives), and flash drives.

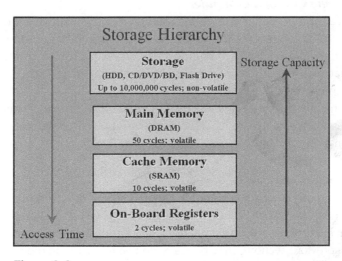

Figure 9-3:
Computer access time and storage capacity tradeoff.

The main memory has a much lower storage capacity than an HDD; for example, a typical computer may have only 4 GBytes of DRAM (G = Giga, or 10^9), while that same computer may have 1 TBytes of HDD storage (T = Tera, or 10^{12}). But the DRAM is almost 1,000,000 times faster than the HDD! And as fascinating as the operation of these devices is (DRAM, HDDs), that is a topic for the next chapter. This chapter is focused on the computer.

So, basically, the CPU is always just doing one of the following things:

1. Getting data from memory
2. Putting data into memory
3. Getting input from one of the input devices
4. Sending output to one of the output devices
5. Operating on, or processing, the data

The fact that a computer appears to be capable of doing all these things at once is simply due to the fact that a computer can do each of these things very quickly. A computer can do each of the above things once in a matter of a few nanoseconds (nano = 10^{-9}), which means it can do all of these things hundreds of millions of times every second!

Inside the CPU of a Computer

Since most of the portions of a computer described above are basically familiar to most readers of this book, the preceding portion may seem quite simple—and it is! But most readers are probably not deeply familiar with the insides of a CPU, so its contents may seem to be incomprehensible.

Thankfully, this is not at all the case. A typical CPU has only six basic parts, and all of them are readily comprehensible. This organization is shown in Figure 9-4.

PROGRAM (IN MEMORY)

A computer program is stored in the memory of the computer and is nothing more than a list of instructions for the CPU to perform. These instructions are very limited in their basic nature (as described in the previous section); they basically do five groups of things:

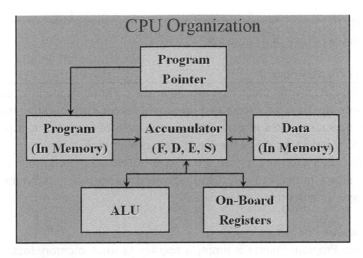

Figure 9-4:
The organization of a typical CPU.

1. Move data (to or from memory; to or from the CPU; from an input device; to an output device; between specific storage locations within the CPU itself).
2. Operate on the data in the CPU in a specific way (add, subtract, multiply, divide, AND, OR, and a few other more arcane types of operations).
3. Change the Program Pointer (which means to change the flow of the program). These instructions also include the conditional operators, which will be described in more detail later.
4. Count (usually how many times a particular operation has been done).
5. Compare two values and set a flag according to the results of the compare.

Depending on the actual CPU and its specific design, there may be other unusual types of instructions, but the above five groups of instructions are all that are necessary to perform the many amazing and complicated things that a computer does. Given enough of these instructions (which is the software of a computer) a computer can be made to perform almost any task, so these instructions form the heart of the versatility of a computer.

DATA (IN MEMORY)

The data for the program is usually in the computer's memory, and is both the data to be operated on as well as the data resulting from the operations already performed. In the case of a spreadsheet program, the data would be the actual spreadsheet itself.

The data and the program are always stored in binary format, and this binary format can be used to represent numbers, letters, symbols, images, or any other type of data.

PROGRAM POINTER

The Program Pointer is simply a register (a small memory location inside the CPU) which keeps track of which instruction needs to be executed next. The order in which the program instructions are executed is very important; just like with a recipe, the wrong order will produce the wrong results. So the job of the Program Pointer is very simple—just keep track of where the CPU is in the instruction list.

ON-BOARD REGISTERS

As mentioned in the preceding paragraph, a register is simply a small memory location inside the CPU. Most CPUs have at least 8 registers, but seldom are there more than 64. This is a VERY small amount of memory, but the main purpose of these registers is to simply allow the CPU to have places to store the data it needs again very soon. Because these registers are right inside the CPU, they can be accessed very quickly.

ALU (ARITHMETIC LOGIC UNIT)

As described by its name, this is the place in the CPU where data is actually operated on, where the addition, subtraction, multiplication, division, or logic operations are performed. A CPU can only operate on two values at a time, not more. If a program needs to operate on more than two values, then successive operations are necessary.

An ALU is made up of some very specific types of electronic circuits which perform these operations on binary numbers only. If decimal numbers are being operated on (which is usually the case), then the program has to convert the decimal numbers first into binary, then operate on the binary numbers, and then convert them back to decimal after all the necessary operations are done.

ACCUMULATOR

The last functional block to be described is the Accumulator, which is where the results of each operation are temporarily stored—and thus the name! The convenience of the accumulator is that in most CPUs, the value stored in it can be operated on with any register or memory location, and the results of that operation are again stored in the Accumulator, which allows successive operations to be performed iteratively, which simplifies the programming.

In the block diagram of Figure 9-4, the Accumulator is shown with (F, D, E, S)—this is a reminder of what the CPU is ALWAYS doing: Fetch, Decode, Execute, Store. After it does this instruction cycle once, it does it again, and again—as long as the computer is turned on, it is doing this, CONSTANTLY. The CPU fetches an instruction from the program (in memory—pointed to by the Program Counter), decodes it (meaning it prepares the hardware of the CPU to complete the desired instruction), executes the instruction, and then stores the result in the location specified.

But what does a computer do when it has nothing to do, if it is always doing this F, D, E, S instruction cycle? If the CPU has nothing to do, it usually goes to a little Idle program, in which it performs some meaningless operation over and over (such as count up or down), or actually performs an instruction known as NOP—No Operation. This instruction does nothing, after which the CPU returns to that same instruction and does nothing again—millions of times a second—waiting for something to do. In between each iteration of this NOP, it may check whether there is anything new to do, or it may simply wait for an operation to interrupt the continuous iteration of this very empty process. But a CPU must ALWAYS be executing an instruction. Which brings us back to what we said at the start of this chapter: a computer is an electronic instruction executor.

Computer Programs

The native instructions of a CPU are very useful, but somewhat arcane, and it takes many thousands of them, in exactly the right order, to accomplish something meaningful. And writing such programs is extremely tedious. But for the first several years after the computer became widely used, this was really the only way to program them. These native instructions are different for every type of CPU, and are known as *assembly language* instructions.

It wasn't long, however, until higher-level languages were invented, and these improved the effectiveness of programmers significantly. Some of the early high-level languages (HLLs) were COBOL (COmmon Business-Oriented Language) and FORTRAN (FORmula TRANslator). These HLLs allowed programmers to express the desired instructions in a way that was much easier to express, to read, and to comprehend.

But most computers cannot execute an HLL directly—the instructions are not in a form that the CPU can fetch and decode. So a special program has to be written called a *compiler*, which takes this HLL program and converts it into a program in the native language of the CPU.

Yet even with HLLs, the task of programming a computer remains one of the greatest challenges of our day, especially when it comes to highly complex programs. For example, some of the most popular programs for today's personal computers (word processing or games) easily consist of millions of lines of instructions in some HLL, which in turn are translated into tens of millions of lines of instructions in the native instructions of the CPU.

So why does a certain word-processing program have 84 fonts to choose from? Because some programmers wrote the software that made it have these 84 fonts to choose from. And why does that same word-processing program cause a drop-down menu to appear when the mouse moves over an icon? Again, because some programmers wrote the software to make it do that. And it takes thousands of instructions, just to do each of these rather simple operations! What that does for me is give me a great deal of gratitude for the tens of thousands of people who have spent millions of collective hours writing software which has made so many tasks so much easier than they ever were before. Without a doubt, were it not for the word-processing program

I'm using at this very moment, this book would never have been written. This author learned to type on a typewriter, and also learned to write in cursive, but neither is a practical way to write a book when compared to the powerful tool that a word processor is.

How Does a Computer Sense Something Outside It?

One of the things that makes a computer seem intelligent (and they AREN'T intelligent) is the fact that it can respond to outside input, and it can change what it does based on that input. One of the simplest examples of this is how a computer senses the stroke of a key on a keyboard.

There are two main mechanisms by which a computer responds to external input. The first is by the use of a conditional instruction; the second is by the use of an interrupt. An analogy with how we humans make decisions would probably be helpful.

CONDITIONAL INSTRUCTIONS

A conditional instruction is akin to how we decide whether to wear a coat: we check the outside thermometer, and IF it's colder than whatever our personal "cold threshold" is, we go to the closet and retrieve our coat. In other words, we execute the conditional instruction "retrieve coat from closet" *only* on the condition that it's cold outside.

Most computers have many conditional instructions, and they can be used almost anywhere in a program. These instructions are only executed if the condition is met, and that condition is usually something very simple, such as a key on a keyboard being pressed. For this example of a key on a keyboard, the program would instruct the CPU to check the keyboard input and see if something is there. Only if there IS something there does the CPU retrieve the key that was pressed; otherwise, it does nothing.

INTERRUPTS

An interrupt is akin to how we respond to the input that our house is on fire. Regardless of what else is happening in our lives, we stop what we're doing and respond. Everything else is less important until we deal appropriately with that emergency.

Most computers have several types of interrupts, and a keyboard interrupt is one of these specific types. An interrupt is a change in the flow of a program, which is initiated by external input. Each type of interrupt immediately causes the computer to change from the next instruction it normally would have executed, to instead executing the set of instructions written to service that specific type of interrupt. By their very nature, interrupts are serviced very quickly, and as soon as they are serviced the CPU is returned to the very instruction it was about to execute before the interrupt occurred.

How Does a Computer Store and Find a File on a Hard Drive?

The process by which this amazing miracle takes place is detailed in Chapter 10 on Computer Storage. Suffice it to say that this happens due to high-level details being broken down into lower-level details, which details are in turn broken down into even more details. This continues until the final level of detail is reached: the instructions actually executed by the computer.

At the highest level, the computer simply keeps track of where on the hard drive each file is kept. This information is also stored on the hard drive. So when the computer needs to retrieve a file, the first place it looks is in the directory on the hard drive. This directory tells the computer where to find the file, after which it then goes to that location on the hard drive and retrieves the file.

How Does Software Interact with Hardware?

This question is insightful by its very nature, and is often asked by people first learning about computers. The answer to this question is closely related to the answer to the question about how a computer senses and responds to something outside of it.

We're all basically familiar with how a switch causes a light to come on or go off—it is in the path of the electricity going to the light, and unless that path is completed by closing the switch, the electricity cannot cause the light to come on. A light switch is a classic example of a hardware switch.

Software also can have switches, commonly referred to as "flags", but they don't force something on or off UNLESS the program tells the computer to do so. It is very easy to store in the CPU registers the status of things going on inside and outside the computer, and it is also very easy to write programs which use conditional instructions, based on these flags. And so common is this need that there is a dedicated register in a CPU known as a Status Register, and there are conditional instructions which are only executed depending on certain bits (flags) in the Status Register.

The bits in the Status Register can be set based on external input, or based on the results of previous instructions. By writing programs which use conditional instructions based on the bits in the Status Register, the computer can be made to respond to external inputs, which thus allows the software to respond to the hardware.

What Are the Applications of Computers Today?

This may seem a silly question, given the pervasiveness of the computer in modern society. So, it is not my intent to create some all-exhaustive list of everything a computer has ever been used to accomplish. Not only would such a list exhaust the reader, but it would surely also exhaust the author!

But it has been my experience that there are many things that computers do today that some people are unaware of, and listing a few of these could indeed be fun and perhaps even insightful.

EMBEDDED COMPUTERS

This category of computer applications is surely the most diverse and pervasive. Embedded computers are computers that have no keyboard, mouse, or monitor, but are nevertheless an integral part of today's devices. A classic example would be the automobile (see Chapter 16). Most of today's vehicles have approximately 25 micro-controllers, which are simply a small kind of computer. A typical one would monitor all the variables of engine performance (outside air temperature, air pressure, humidity, engine temperature, engine speed, etc.) and would then cause the fuel injectors and timing to respond accordingly, allowing the engine to achieve near-optimal performance.

Computers are inside nearly all modern stereos, CD & DVD players, MP3 players, cameras, camcorders, cell phones, and almost everything electronic. But they are also in most washers, dryers, refrigerators, stoves, microwave ovens, mixers, sewing machines, and other kitchen appliances. A very simple kind of computer is even found in self-flushing toilets and urinals. And a few sports shoes have become so high-tech that they also include a small computer to optimize how they respond to their owner.

Elevators have computers; golf carts have computers; even some football helmets have a computer inside! Hospital beds have a computer; some watches have computers; some roads, bridges and buildings have sensors and computers; many electric tools have a computer. Furnaces, air conditioners, compressors—most of these have a computer.

And so it goes; so pervasive is the computer in modern society that we almost assume that anything electronic or electric has a computer in it, and this is generally true!

Chapter Take-Aways

Only a few decades ago, computers were non-existent. In the 1940s, they started to appear, but they were extremely large, expensive, power hungry, and limited in their usefulness. No one's daily life had anything to do with computers.

Today is an incredible contrast. In first- and second-world countries, there is hardly an activity in which people engage where they do not interact with a computer (whether they know it or not). The examples in the preceding section should serve to illustrate this. Thus it is little surprise that if all computers were to somehow cease to function, our society would quickly grind to a halt. We have become incredibly dependent on the computer, and while it has surely added to the complexity of our life, it has also dramatically enabled new activities and insights, greatly enriching our lives.

Conclusion

There are many marvels of electronics discussed in this book, and all of them are truly remarkable. But in this author's opinion, none of them can compare to the versatility, pervasiveness, and utility of the computer. And since Moore's Law (see Chapter 4) has made them ever smaller, more powerful, less expensive, and more useful, they just continue to find more uses. And it does not appear that this is likely to decrease in the coming years. In many ways, electronics has become quite centered around this prodigy, and justifiably so.

References

1. Leavitt, David, "The Man Who Knew Too Much: Alan Turing and the Invention of the Computer", Atlas, 2006.
2. Nardo, Don, "Computers: Mechanical Minds (Encyclopedia of Discovery and Invention)", Lucent Books, 1990.

CHAPTER 10
WHERE DO YOU PUT IT ALL?–
COMPUTER STORAGE

The Need for Computer Storage

Since the invention of the computer, it has always been the case that the actual programs which are executed by the computer, and the data on which the CPU operates, are stored external to the CPU itself. Originally, this was because the technology necessary to store programs and data was quite different from the technology of the CPU. The invention of the integrated circuit changed that, but it wasn't until several years after the invention of the microprocessor that the CPU itself actually contained a significant amount of storage. But even today, regardless of the rather significant sizes of storage available on the CPU itself, it is very rarely enough, so most programs and data for computers are stored elsewhere.

Desirable Attributes for Computer Storage

At first glance, this section seems rather obvious and thus unnecessary—we want computer storage to store our programs and data. What other attributes are there? Well, it turns out that there are many desired attributes that begin to fall out when we learn what's out there today—no type of computer storage is ideal, even in today's modern world. Each option for storing data has its strengths and weaknesses. Since the importance of these attributes varies by the application, they cannot easily be ranked by importance, so we will order them alphabetically. These attributes will be discussed in more detail below, and include the following:

1. Capacity: large
2. Cost: inexpensive
3. Nonvolatile (meaning that the data remains even after the power is turned off)
4. Power consumption: low
5. Read/write cycles, number of: infinite

6. Size: small
7. Speed: fast

This list pretty well covers the attributes we would recognize as important, at least at first. It turns out that there are others, but they only make sense when we first understand why we even care about them, and that's yet to come.

CAPACITY: LARGE

The ideal data storage device would hold as much data as we could possibly want to put on it, plus a reasonable margin more. Early storage devices were considered large if they could hold a few MB (Megabytes); today's large devices hold about 1 million times more than MB—upwards of tens of TB (Terabytes), which is tens of trillions (1,000,000,000,000) of bytes. And sometimes we still wish they could hold more!

COST: INEXPENSIVE

Cost is essentially always a consideration, and usually one of the most important. Indeed, the dominant storage technologies today are dominant for that very reason—they are able to store the most data at the lowest cost. Unfortunately, they bring with them a few undesirable characteristics, but more about that later.

NONVOLATILE

The first computer storage was nonvolatile, meaning that even if the power was turned off, the data remained. Once the power was turned back on, the data could be read again. It has always been desirable that computer storage be nonvolatile. But the majority of today's primary storage (meaning the first place the computer looks for the data) is volatile, and we've all experienced the frustration of that. If you don't save what you're working on and the power suddenly goes off, you lose everything you haven't saved.

The first question this usually raises is, "Why?" Why did we give up our preference for nonvolatile storage? The answer is the first desirable characteristic: cost; more about this later.

POWER CONSUMPTION: LOW

The ideal computer storage would not require ANY power, but we've all become very accustomed to accepting the fact that this is not really possible. But at least it should not take much power. This is true from a "green" perspective, as well as from a practical perspective. A technology which does not consume much power is much more practical than one that consumes a lot of power.

READ/WRITE CYCLES, NUMBER OF: INFINITE

Computer storage is continuously reused. Data stored in one location today will be stored in another location tomorrow, and the previous location will be used by other data. Therefore it is desirable that the computer be able to change the data (rewrite) as often as necessary, without decreasing the lifetime of the storage technology.

Again, this seems quite obvious, and the first computer storage had no problem with this; it could be rewritten as many times as needed, even an infinite number of times. And this was true for the second main type of primary computer storage. But as various types of memory were developed, eventually one was developed which was *non*volatile! But alas, the disadvantage it had was that it could only be rewritten a limited number of times. Again, more about this later.

SIZE: SMALL

This characteristic is also very obvious, and has been a major driving force behind most improvements in computer storage technology. Everyone wants to be able to store more data in less space. In order for a new storage technology to be viable, this characteristic is essential.

SPEED: FAST

This one is also quite obvious; everyone prefers a fast computer to a slow one. And accomplishing this has not been easy, as will be obvious from the efforts described in the following sections.

What's a Computer To Do?

One very fundamental fact about computers needs to be understood: they are data consumers. A computer ALWAYS needs data. As long as they are powered on, they are ALWAYS looking for something to do, and that something involves data. As soon as a computer completes one instruction, it immediately gets the next instruction and begins working on that one and its associated data. It does this millions and even billions of times a second!

So what does a computer do when there's nothing to do? Most of our personal computers are not busy most of the time. So when they're idle, are they still getting instructions and executing them? Yes. And the instruction they execute when they're idle: a small program, called the System Idle Process. Figure 10-1 was taken from the author's computer, on which this book was written. It is a list of all the programs being run on the author's computer at that time. It shows 42 processes (programs) being run on the computer, including familiar ones such as Firefox, Thunderbird and WINWORD (MicroSoft Word—which is being used to write this book). At the moment when this Windows Task Manager snapshot was taken, the System Idle Process (the last one on the list) was taking up 95% of the CPU time. In other words, the computer had very little to do (only occupying 5% of its time), and the rest of its time was being spent executing this System Idle Process. The moral of the story is that a computer ALWAYS has to have something to do, even if that is doing nothing.

How does a computer know what to do? The operating system, such as Windows, (which is also a program) keeps track of that, and points the computer to where the instructions and the data are. And since the data cannot all be stored on the CPU, most of the time the data is being fetched from somewhere else.

This leads to a storage hierarchy, as depicted in Figure 10-2. At the bottom of this hierarchy is storage which can be accessed very quickly—only a few cycles of the system clock, which means only a few nanoseconds (billionths of a second). But it doesn't have much capacity—only a few hundred bytes. And it's volatile—if the power goes off, the data is *gone*.

At the top of this hierarchy is rotating mass storage, such as hard-disk drives and optical disc drives. They hold a great deal of data, from Gigabytes to even Terabytes, but it takes a relatively LONG time for the data to be accessed—from several milliseconds (thousandths of a second) to even hundreds of milliseconds. While this may seem fast

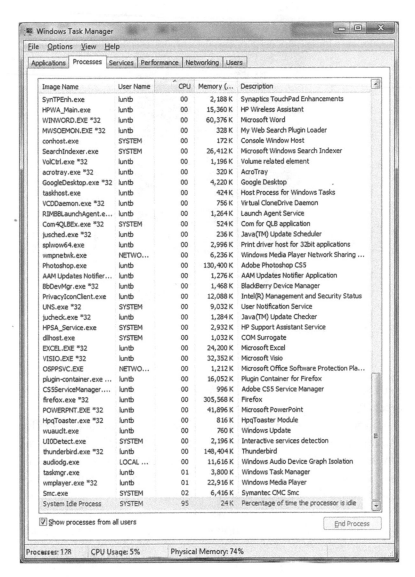

Figure 10-1:
Processes (programs) being executed on the author's computer.

135

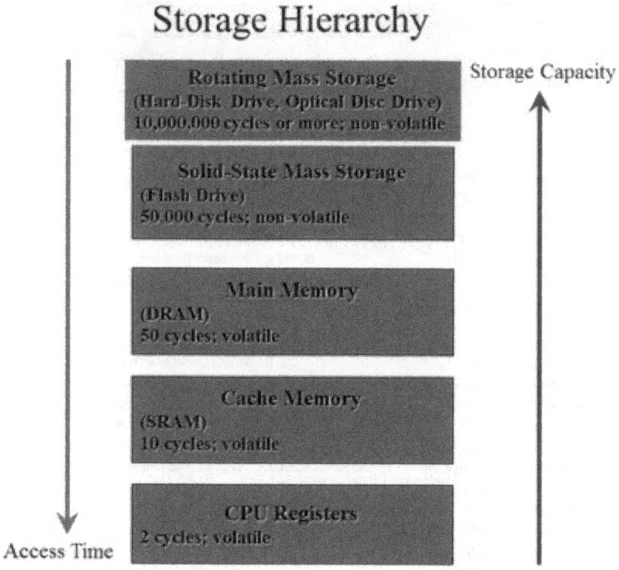

Storage Hierarchy

Figure 10-2:
Computer storage hierarchy.

to us, keep in mind that this is *millions* of times longer than it takes to access the fastest storage.

But the cost per byte is also a major factor in this storage hierarchy, and the least expensive storage, in terms of cost per byte, is also the slowest—rotating mass storage. If a computer had to go to rotating mass storage to get every instruction, that computer would be more than a *million* times slower than an ordinary computer. And it is also true that one of the easiest ways to speed up a computer (if it is low on main memory) is to supply it with more main memory, simply because then most instructions can be found in main memory, instead of having to go to the hard-disk drive (HDD), which is *so many* times slower.

One final point about all this fetching of instructions. Yes, a computer is constantly executing instructions, and therefore constantly going out to memory to fetch instructions. But it is also constantly executing these instructions on data of some sort, and this data also resides in memory or on the HDD. So a computer is constantly (in fact, billions of times every second) going out to memory (cache or main) to find instructions or data. If the needed instruction or data is not in memory, then the computer goes out to the HDD or other mass storage to obtain the needed information.

A word about computer boot-up is also helpful here. When a computer is first turned on, all of the instructions and data that were previously resident in memory are gone. In order for the computer to know what to do, it must first load the operating system into memory, after which it is then ready to load whatever application programs are needed. This accessing of the operating system, followed by the necessary application programs, is most of what takes a computer so long to boot up—it is simply fetching the information from mass storage and loading it into memory.

Historical Development of Today's Options

Where we are today in terms of our options for storing digital data has much to do with where things have been historically; the rest has to do with the desirable attributes discussed in the first section. Technology has constantly been battling to improve digital data storage in one or more of these desirable attributes. And as improvements have been made in one attribute, it has usually had repercussions (both good and bad) in other attributes. This section will briefly discuss the historical development of these storage options.

MAGNETIC CORE MEMORY

As an example of this, we should probably start the discussion by describing a type of digital data storage which is now essentially unused: magnetic core memory. Most of us are familiar with the fact that iron can be magnetized; it is also true that iron can be magnetized in either of the two magnetic polarities characteristic of magnets. Thus, the fact that there are two magnetic polarities available allows us to store both 0s and 1s magnetically by simply defining one polarity as a 1, the other as a 0. And one very nice characteristic of this type of storage is that it is nonvolatile; once we polarize the iron, it remains in that polarity until we force it to reverse. Another nice characteristic is that this process can be reversed an infinite number of times.

Figure 10-3:
Core memory, from the very early days of the computer. This example has 16,384 bits of memory, or 2 kBytes. It is about 4" on each side. In the small magnified region in the lower right-hand quadrant, the tiny cores are seen standing vertically at 45° angles to the quadrants, and with vertical, horizontal, and diagonal wires passing through each core. Just imagine what it would take to have 1 GBytes of this memory!

But, as always, there were several drawbacks to this technology, the first being the size of the bits. It was quickly learned that the optimal shape of these magnetized pieces of iron would be the shape of a ring (donuts, or magnetic cores, as they were called). But these magnetic donuts had to have wires going through them to allow them to be magnetized and to read out the data. Figure 10-3 shows what a plane of these magnetic cores looked like. It can be seen from Figure 10-3 that shrinking the size, weight, cost, and power requirements of this type of storage is very difficult to do.

SOLID-STATE MEMORY

These problems with magnetic core memory soon led to two of the technologies dominant today: solid-state storage and magnetic tape and hard disc drives. Solid-state storage (which means using transistors) was introduced in the early 1960s, and soon became cost-competitive with magnetic core memory. It was also much faster, required less power, and was *much* smaller. It was obvious that these advantages were sufficient to allow it to begin to replace magnetic core memory. It came in two basic flavors: SRAM (Static Random-Access Memory) and DRAM (Dynamic RAM).

But there was, and still is, one major drawback to solid-state memory: it is volatile. This meant that a computer had to always be left on (which was usually the case with early computers anyway). But even if the computer was never turned off, it was still highly vulnerable to power glitches. Clearly, something had to be done to provide a nonvolatile way to store computer data if solid-state memory was to be viable.

MAGNETIC TAPE AND DISCS

Storing voice and music on magnetic tapes goes back to even before the early days of the computer, so it was obvious that the advantages provided by magnetic tape could be adapted to storing digital data. This led to the development of ½-inch magnetic and tape drives in the 1950s. These tape drives (see Figure 10-4) soon became icons of the computer realm.

Half-inch tape drives solved the problem of volatility, but they had their own problems, one of which was "sequential access". If you needed a file which was at the end of the tape, you just had to wait until the tape could be moved to that location, which could take several minutes.

Soon after ½-inch tape became widely used, a new technology was invented which stored data on flat, spinning magnetic discs; this technology was the birth of today's hard-disc drive (HDD) industry. This storage technology, like magnetic tape, was also nonvolatile, and it featured "random access", as opposed to the sequential access of magnetic tape. With spinning discs, the longest wait was only until the disc

Figure 10-4:
Several ½-inch tape drives from the 1950s and 1960s.

spun around one more time, which was on the order of hundredths of a second—MUCH faster than the sequential access of ½-inch tape.

Today, ½-inch tape and HDDs are still with us, although much reduced in physical size, power requirements and cost, and much improved in speed and capacity. They also provide nonvolatility and an infinite number of read/write cycles.

OPTICAL DISC STORAGE

Ever since the invention of the laser, people have envisioned using these highly-focused beams of light to store data, and many products were introduced which attempted to commercialize this technology.

The first digital optical format to be widely accepted was the Compact Disc (CD), which was soon followed by recordable versions (CD-R and CD+R) and even re-recordable versions (CD-RW and CD-MO). These storage technologies provided relatively small size and low power consumption, low cost, and reasonable cost and capacity. Less than a decade later, the CD was succeeded by the Digital Versatile Disc (DVD), which was also soon expanded to include recordable and re-recordable versions (DVD-R, DVD+R, DVD+RW). More recently, the Blu-ray disc (BD) has also become widely accepted, and is also available in recordable and re-recordable versions (BD-R, BD-RE). All optical discs feature non-volatility; only the re-recordable versions can be erased and re-written. They provide excellent cost, reasonably small size, and reasonable speed.

FLASH MEMORY

But as viable as the above technologies were, they suffer from one major drawback: their access time is in the range of milliseconds (thousandths of a second). While this is considered reasonable for mass storage which is nonvolatile, it has been discussed previously that it is still far too slow to serve as the main memory of computers. Storage needs to be thousands or even millions of times faster than this in order to be viable for the main memory.

Solid-state memory meets the need for speed, but has historically suffered from the major drawback of being volatile. So over the course of the past 40 years, several non-volatile versions of solid-state memory have been invented, each with their own strengths and weaknesses. These include ROM (Read-Only Memory), PROM (Programmable ROM), EPROM (Erasable PROM), EEPROM (Electrically Erasable PROM), and Flash. Of these, Flash is the most recent and the most widely accepted, and is now widely used in Flash drives (also known as "thumb" drives or "memory sticks") and in storage for digital cameras. It is also available as a pseudo-HDD, in that it uses Flash memory but acts like a HDD as far as the computer is concerned; this format is known as SSDs, or Solid-State Drives.

SUMMARY OF TODAY'S OPTIONS

After this brief historical review, what we have as today's options for computer data storage falls into three main categories: magnetic (including HDDs and magnetic tape), optical (including CD, DVD, and BD), and solid state (composed primarily of DRAM, SRAM, and Flash). Each of these technologies will be further described in the following sections.

Basic Operation of Solid-State Memory

All solid-state memory uses transistors to store data. However, the construction of the cell (the lowest unit of storage in solid-state memory; 1 cell = 1 bit) differs depending on which type of solid-state memory we're talking about. The types available today can be classified as DRAM, SRAM, and Flash memory. They feature access times in the range of a few nanoseconds (billionths of a second) to a few hundred nanoseconds. While this is still slower than the computer would prefer (for instance, a 4 GHz microprocessor needs an instruction typically every 0.25 ns), it is still *much* faster than anything in magnetic or optical storage.

DRAM

The first type of solid-state memory we will describe came out in the late 1960s, and used only one transistor and one additional element (a capacitor) for each cell. It reduced the size of a cell to what was considered the smallest it could be. It can be thought of as a faucet with a small water balloon attached. If we turn the faucet on for a short time, we can fill up the water balloon. Later, we can come back and see if the water balloon is full or empty. If full, we interpret it to mean that a 1 is stored there. If the water balloon is empty (it never was filled), we interpret that to mean that a 0 is stored there.

In real solid-state circuits, the function of the faucet is performed by a transistor; the function of the water balloon is performed by a capacitor. The transistor turns on and allows the capacitor to be filled with charge (electricity). Later, we can come back and check whether the capacitor is filled or empty, thus finding out what data is stored there.

But there is a very big problem with real transistors and real capacitors—they leak. The transistor faucet, when turned off, still allows some of the charge stored in the capacitor to leak back through the faucet, which will eventually empty the capacitor. Additionally, the capacitor is a leaky water balloon, and even if the faucet were perfect, it would

Figure 10-5:
A DRAM cell is much like a tiny water balloon being filled from a faucet. The transistor performs the function of the faucet, and the capacitor performs the function of the water balloon.

eventually leak all the charge away. And the worst part of this is that together, these leaky devices are so bad that the charge in the capacitor water balloon leaks away in about 1 ms (1/1000th of a second)! So what good is a leaky faucet and a leaky water balloon as a storage element? It seems entirely impractical!

But the reality is that 1 ms is a relatively long time to a computer; it can execute millions of instructions in that time period! So, we just supply these DRAM memory chips with some special electronic circuits which spend ALL their time going to each memory cell, one at a time, checking what's been stored there, and filling the little water balloon back up if it's getting a little low. This process, known as "refreshing", is what makes DRAM dynamic—it's always changing (leaking)! But because the refresh circuit keeps ahead of the leaking, a constant state of storage is maintained—as long as there's power turned on!

Figure 10-6:
The main memory of modern computers is usually packaged in SIMMs (Single-In-Line Memory Modules) or similar small circuit boards. This one is about 6" (15 cm) long by 1" (2.5 cm) high.

Because DRAM only requires one transistor and one capacitor per cell (bit), and because it has been improved continuously for decades, it is the most dense of the three types of solid-state memory, meaning it has the most bits in a single chip. This also makes it the least expensive, which is why DRAM is used for the main memory of a computer. Also, because the cells for DRAM are so small, it requires relatively little power. And there is no limit to the number of times we can change the contents of DRAM. Figure 10-6 is an example of DRAM, packaged in a SIMM (Single In-line Memory Module), as you would probably find it when you buy it in a store or online.

SRAM

The second type of solid-state memory we will describe also came out in the late 1960s, but its main purpose was to be fast. It used six transistors per cell (see Figure 10-7), and was therefore not as dense as DRAM (which also means it was more expensive.) But because it could use these transistors in a reinforcing way, it was static (meaning it did NOT change with time) and it was fast. SRAM has a typical access time between 5 to 10 times faster than DRAM.

The basic storage cell of SRAM works in what is known as a bistable circuit—it is very stable in either state, but something must be done to make it change from one state to another. It is very much like a mechanical light switch; it will stay up or down indefinitely, but you

144

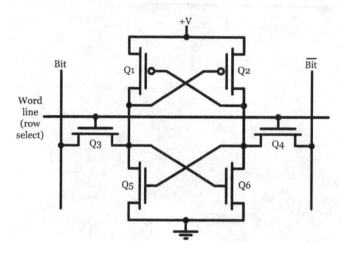

Figure 10-7:
A schematic diagram of an SRAM storage cell, using 6 transistors (Q1–Q6) to store 1 bit.

must move it to make it change. The change is caused by circuits which store the data in each cell.

As long as there is power to the SRAM, the data contents remain stored indefinitely. But when power is lost, the data in SRAM becomes much more like a coin on a table, but the table has been tipped over. Before power is lost, the coin is either heads-up or tails-up, and will remain in that state indefinitely. But when the table is tipped over, everything is lost. When we turn the power back on, we right the table and restore the coin to the surface of the table, but we no longer have any idea if the coin is heads-up or tails-up.

So, the strength of SRAM is its speed, which is why it is used for *cache* memory. It doesn't hold as much data per chip as DRAM, but it is much faster. It requires a bit more power than DRAM, but less than optical or magnetic storage (while reading or writing). And there is no limit to the number of times we can change the contents of SRAM. Figure 10-8 is an example of SRAM as you might find it on a computer motherboard.

Figure 10-8:
SRAM as found on the motherboard of a desktop computer.

FLASH

The last type of solid-state memory we will describe is the most recent, having been introduced in the early 1990s. But its history goes much further back, as mentioned in the previous section. Its predecessors include ROM, PROM, EPROM, and EEPROM. Indeed, the full name of this type of solid-state memory is actually Flash EPROM.

All of these types of memory were developed to try to address the greatest weakness of DRAM and SRAM—their volatility. All of these types of memory are nonvolatile, but this characteristic came at a very high cost for the first type—ROM could not be written to, thus its name, Read-Only Memory. Further advances allowed them to be written to once (Programmable ROM). The next advance allowed them to be fully erased and rewritten, a limited number of times (Erasable PROM), but they had to be removed from the computer to do this. This problem was solved in the next version, the Electrically-Erasable PROM—it could be erased while still in the computer. But this could only be done a limited number of times (thousands to tens of thousands of times).

The most recent offspring of this line of development is Flash EPROM. It can be erased while still in the computer (or digital camera, or whatever it's installed in), it is nonvolatile, and this version can be erased many millions of times. It has become the overwhelming favorite type of nonvolatile solid-state memory.

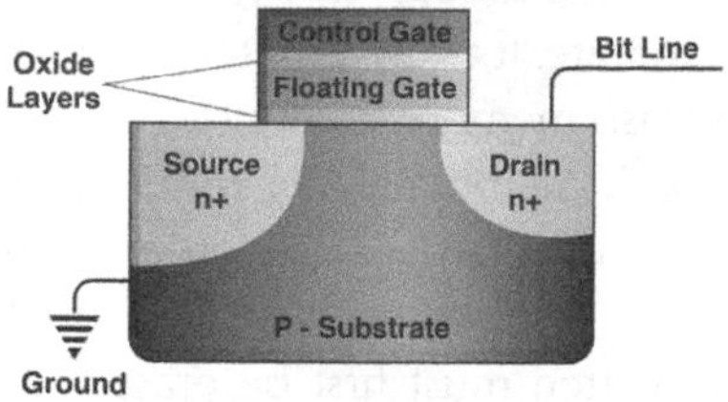

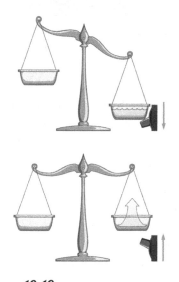

Figure 10-9:
Flash memory works by storing charge on a floating gate of a transistor.

The basic storage element of Flash memory is a transistor with a floating gate (see Figure 10-9). It operates much like the scale-switch combination shown in Figure 10-10. If we fill the cup on the right scale with water, the scale will tip in that direction, which will cause the spring-return switch to turn on. As long as there is water in the cup on the right side of the scale, the switch will remain pushed down, and thus on. If the water were to evaporate or otherwise leak away, the scale would tip in the other direction, causing the switch to return to the off position.

In actual Flash memory cells, the floating gate operates as the cup and stores charge (roughly the same as electricity); this charge causes the transistor to turn on and stay on, as long as the charge remains on the floating gate. But as the insulating material surrounding the floating gate is not perfect, the charge does

Figure 10-10:
Storage of data in a Flash memory cell operates much like storing water in a cup on a scale, which is over a spring-return switch.

leak off, although slowly—just as the water in the cup would slowly evaporate. It takes about 10-12 years to begin losing some of the data in Flash memory.

Flash memory is not quite as fast as one might think, at least for the writing process. In order to write new data to Flash memory, the block (usually 512 bytes) of data into which the new data will be written must first be erased—all bits in the block. Then the new data is written to each cell in the block. This takes about 10 μs (10 millionths of a second), which is much faster than a hard drive or an optical disc drive, but it is much slower than SRAM or DRAM.

Although Flash memory is a bit slow on the write time, it is rather quick on the access (read) time, in the same range as DRAM—30 to 50 ns.

In summary, Flash memory is about as inexpensive as DRAM, is nonvolatile, can be

Figure 10-11:
Examples of Flash memory sticks. Each of these contains 2 GBytes, yet is quite small in size.

rewritten millions of times, requires relatively little power consumption, is very small in size (see Figure 10-11), and is very good in speed. It is presently the dominant form of solid-state nonvolatile memory.

SUMMARY OF SOLID-STATE MEMORY

Table 10-1 provides a summary of the characteristics of the three main types of solid-state memory, listed alphabetically. All three types are widely used in today's amazing array of electronic devices. This class of solid-state storage is the only class of storage fast enough to be used for the main memory or cache memory of a computer. The other two classes of storage to be discussed next (magnetic and optical storage) are much slower, but have the big advantage of being nonvolatile.

Memory Type	Write Speed (ns)	Read Speed (ns)	Volatile or Nonvolatile	# of Read /Write Cycles	Comparative Size
DRAM	30–50	30–50	Volatile	Infinite	Very small
Flash	10,000	30–50	Nonvolatile	>10,000,000	Very small
SRAM	3–5	3–5	Volatile	Infinite	Small

Table 10-1: Characteristics of the main types of solid-state memory.

On the topic of size, it might be fun to learn just how much room it takes to store a single bit in DRAM, SRAM, and Flash memory. Over the last 50+ years, the size of transistors, and thus of memory cell sizes, has continued to drop. Today's DRAM chips can easily be purchased in quantities of 16 Gbits, which means there are over 16,000,000,000 transistors in an area of about 0.4" by 0.4" (10 by 10 mm) of silicon—which works out to a cell size of about 3.16 µinches (80.3 nm) square. For comparison, a human hair is about 3,000 µinches (75 µm) wide, which means we could fit about 933 DRAM cells in the width of a single hair! The size of a DRAM cell is also in the range of many viruses—only visible to scanning electron microscopes!

Flash memory cells are nearly as simple as DRAM cells, and over the years, their size has also continually shrunk, until they are now on a par with the size of DRAM cells.

SRAM cells, which contain 6 times as many transistors as DRAM and Flash memory cells, cannot be shrunk as small, but they are still only about 4 times the size of a DRAM or Flash memory cell. And they are hundreds of times smaller than can be seen with the naked eye.

Basic Operation of Magnetic Storage

Historically, magnetic tape storage was invented before hard-disc drive storage was invented, so that is the order in which we will consider these amazing technologies.

MAGNETIC TAPE—ANALOG

Recording sound on a moving media dates back several decades, clear to the early 1900s, depending on which sources you accept. The first magnetized media was wire, in the form of a tape 0.1" (3 mm) wide and 0.003" (0.08 mm) thick. It moved from one reel to another at the astounding speed of 60 inches (150 cm) per second! This was no small safety hazard, considering that a metal tape of these dimensions, moving at this velocity, could slice through a finger like a butcher's bandsaw. Thankfully, the new field of plastics soon offered another medium to replace this, which consisted of a thin tape of plastic (now usually polyethylene terephthalate or PET—brand-named Mylar™) coated with a very thin layer of iron oxide. This dramatically reduced the cost, weight, and danger involved with magnetic recording.

The recording process involves the carefully controlled interaction between the magnetic tape and the magnetic recording head (see Figure 10-12). Basically, the magnetic field created in the head is imposed on the magnetic tape, which then becomes magnetized in

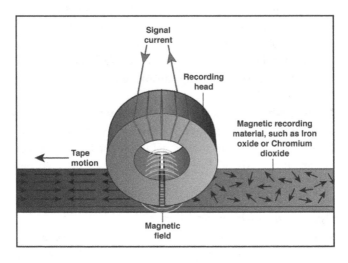

Figure 10-12:
A magnetic recording head interacting magnetically with the magnetic tape.

the polarity and strength of the field created by the head. And since the magnetic tape is moving with time, it becomes a time-based record of the magnetic field created by the head.

The basic function of a magnetic recording head is very simple. If you wind conductive wire (such as copper) around a piece of soft iron and then pass a current through the wire, you create a magnetic field which then magnetizes the iron around which it is wound (see Figure 10-13). If we reverse the direction of the current, we will also reverse the polarity of the magnetic field in the iron, since it can be easily magnetized to either polarity. If we were to bend the iron nail into a loop, with a small gap between the ends, we would then have a magnetic recording head.

The same recording head can be used for reading. As a changing magnetic field passes by the small gap between the ends of the

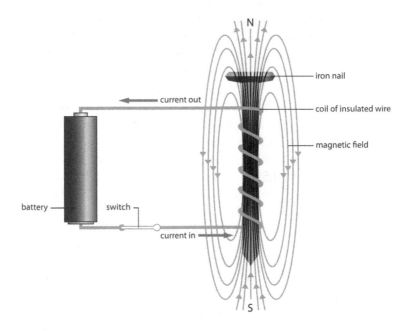

Figure 10-13:
A nail being magnetized by passing electricity through a coil of wire wound around the nail.

recording head, it induces a small voltage in the wire which is wound around the iron. This small voltage represents the changing magnetic field of the magnetic tape as it passes over the read head. After amplification, this voltage signal can drive our speakers and we can hear the sound originally recorded on the magnetic tape.

MAGNETIC TAPE—DIGITAL

Today's digital magnetic recording heads are similar in many ways to this technology just described, although there are also many differences. The materials have been improved and the processes have been improved. And new magnetic effects have made it possible to sense extremely small magnetic fields, which is essential to putting lots of digital information on magnetic tape. Many other advances have dramatically increased the density of magnetic tape storage (the number of bits that can be stored in a given length of $\frac{1}{2}$-inch tape).

For instance, early digital magnetic tape reels were about 10 inches (25.4 cm) in diameter and contained 1400 feet (427 meters) of $\frac{1}{2}$-inch magnetic tape, and recorded at a density of about 267 bits/inch (10.5 bits/mm) on seven parallel tracks, for a total capacity of about 8 million bytes. By comparison, today's $\frac{1}{2}$-inch magnetic tape cartridges are about 5" by 4" (13 cm by 10 cm), contain about 2,620 feet (800 meters) of tape, and are recorded at a density of about 406,400 bits/inch (16,000 bits/mm) on over 1,400 parallel tracks, for a capacity of over 2 Terabytes (2 trillion bytes). While that is an 87% increase in length, it has come with the benefit of a 200-fold increase in the number of tracks and 1,500-fold increase in the linear density, for a combined increase of over 300,000 times in the amount of data that can be stored on a given length of tape.

Although magnetic tape for storing digital information has never heavily penetrated the consumer market (who do you know who personally owns a tape drive for their computer?), it has been solidly entrenched in the role of backup for enterprise data. Thousands of these are sold every year to companies large and small, at prices ranging from only a few hundred dollars to several tens of thousands of dollars, and more. Cartridges for these tape drives range in price from the low teens of dollars to over $100.

Hard-Disc Drives

One of the major drawbacks of tape is what is known as sequential access—if you want a file at the end of the tape, you have to move the tape from the start all the way past the other files on the tape to get to it, which greatly increases the access time. As computer storage needs increased, and especially after the introduction of the volatile SRAM and DRAM memory types for the main computer memory, the need grew for a nonvolatile way to store a large amount of data yet have much faster access to the data than tape could provide.

The most successful technology to meet this demand has undoubtedly been the hard-disc drive (HDD). The very first one was enormous by today's standards—more than twice as big as an ordinary American refrigerator. It took lots of power, had a stack of 50 platters each 24" (61 cm) in diameter, and stored only 5 Mbytes (total!). Data was stored at a linear density of 100 bits/inch (4 bits/mm). While this first HDD was very primitive by today's standards, it was revolutionary in many ways, and established a genre of data storage that persists even today.

Heading up the HDD

Magnetic tape is literally pulled across the magnetic read/write head—it generally stays in contact with the head. In fact, if you know someone with a strong background in this area of storage (including the author), they can tell you stories of magnetic read/write heads which have been heavily worn down over years of use due to the fine sandpaper-like action of the tiny magnetic particles in the magnetic tape rubbing against the head. While this is a problem with magnetic tape, it would make HDDs impossible. The reason is that a HDD is constantly spinning the discs.

Tape drives only move the tape over the head when you actually need to read or write the data. But in order to have fast access to the data on a HDD, the discs must be constantly spinning. If the magnetic read/write heads for HDDs were in contact with the discs, the heads would wear out in only a few days—clearly an unworkable situation.

This problem was solved with the invention of the flying head, a truly innovative technology that required precise control of the flatness

153

of the disc surface and of the geometry of the head. It works because, as the disc spins around, it carries with it a very thin layer of air which is also spinning, though not as fast as the disc. Fluid dynamics allows us to determine the exact velocity and density of this layer of air, as a function of the distance between the disc and the head, the flatness and roughness of the disc, and the air pressure and humidity. So, if we design the front edge of the head (the edge facing toward the rotation of the disc) like a wing, it is possible to make a head that will fly on this thin cushion of air at a fixed distance above the disc surface. While this flying distance decreases the strength of the signal which the head can produce or detect, it solves the problem of HDD head wear—but at a price. In addition to the decreased signal strength produced or detected by the head, there is also the price of the all-too-familiar vulnerability of a HDD to "head crashes".

Flying over the disc—or NOT

The flying head depends on the small cushion of air which rotates with the disc. Any perturbation in the airflow or in the vibration of the disc drive can cause the head to drop closer to the surface of the disc. If the head actually drops to the point of contacting the disc, the usual result is physical damage to the head, the disc, or both. This usually means a permanent loss of data, and often results in a disc drive which is completely nonfunctional.

Much has been done over the years to reduce the likelihood of a head crash, but the problem persists. In part, it persists because the flying height has been dramatically reduced over the years. The first HDD had a flying height of about 800 µinches (20 µm), which is about 1/3 the diameter of a human hair. While this is truly small, it pales in comparison to today's technology, which puts the flying head at a distance of about 0.4 µinches (10 nm)—about 1/7500th the diameter of a human hair! It is truly amazing that this has been done without dramatically increasing the vulnerability to head crashes.

Operation

When a computer needs to write to or read from a HDD, it sends a request for that information to the HDD controller. The HDD controller is usually a circuit board attached to the HDD. The HDD controller then determines where the desired information resides on the HDD, then issues commands that cause the head to move over the correct track (tracks are the concentric rings of data on the disc). Once the head reaches the correct track, it then begins reading the data to determine where the disc is in its rotation. It then waits for the correct location on the disc to come under the head, upon which the read or write operation actually begins. The combination of these actions is known as the access time—the time between when the data is requested and when the HDD locates and begins to provide the data. Originally, this access time was about 600 ms (thousandths of a second); it has since improved to about 5 ms, but this access time can never shrink to the nanoseconds realm of SRAM and DRAM because there is always the mechanical wait time required to physically get to the data—spinning discs take several milliseconds to go around once, even at very high RPMs.

Ever-smaller bits

Over the years, HDDs have improved their storage capacity at an amazing rate, even while shrinking their physical size. Figure 10-14 shows disc sizes over the years, starting with the 24-inch (61 cm) first HDD, passing down to 14 inches (35 cm), 5¼ inches (13.3 cm), 3.5 inches (8.9 cm), and to 2.5 inches (6.35 cm). There are even some smaller than this! Yet with these shrinking sizes, capacities have amazingly not diminished and in fact have increased. Linear densities on HDDs are on a par with those on magnetic tape, mainly due to the incredible advances made in read/write heads, which have become closely tied to semiconductor manufacturing technology. But track densities on HDDs are much greater than on magnetic tape, because the rigid aluminum or glass platters which make up HDDs allow much greater density than can be put on the relatively mushy plastic that makes up magnetic tape. For a comparison, one relatively new tape technology allows about 1,400 tracks at a linear density of about 406,000 bits/inch (16,000 bits/mm), resulting in an areal density of about 1.4 Gbits/inch^2 (17.6 Mbits/mm^2). While that is a LOT of bits in a very tiny amount of space, it is still far behind HDDs, which can pack an areal density of over 400 Gbits/inch^2 (6.2 Gbits/mm^2).

Figure 10-14:
The various platter sizes of HDDs over the years.

This begs the question—which is smaller, the smallest cell in DRAM, or the smallest bit on a HDD? Given the densities discussed above, a DRAM cell is about 3.16 μinches square (80 nm square), while a bit on a HDD is about 1.58 μinches square (40 nm square)—so the magnetic bits are the winner for the smallest!

Overall

Hard-disc drives, which have been with us for several decades now, continue to make improvements in their capacity, their data transfer rate, and their cost. Clearly, these improvements cannot continue forever, but current projections indicate there's still at least another decade left in this technology. They provide random access to the data (as opposed to the sequential access of magnetic tape), the data is nonvolatile, and the size of the drive is reasonably small. Although they sometimes fail catastrophically (which is why we are advised to back up the data on our HDDs), they are ubiquitous when it comes to storing data for computers. Their main drawback is their slow access time, requiring tens of millions of CPU cycles to retrieve the data. Which is why SRAM, DRAM and Flash memory will also be with us for a good while to come!

Basic Operation of Optical Storage

Optical disc drives for storing digital data have been around since shortly after the introduction of the CD (Compact Disc) in the mid-1980s. They have gained wide acceptance for use in audio and video playback, and also for storing digital data from our computers. There are three main types of technologies in each of the three optical disc formats; some of the principle characteristics of each are summarized in Table 10-2. In actuality, there have been MANY other formats and technologies introduced, but the ones listed in Table 10-2 have become, by far, the most dominant, so we'll stick to these.

All three of the main optical disc formats (CD = Compact Disc; DVD = Digital Versatile Disc; BD = Blu-ray Disc) use the exact same size physical disc, which is not an accident. This allows BD drives to be backward-compatible, which means they can be made to read all three formats. The advantage is obvious, so the optical disc storage industry has worked hard to provide this backward compatability.

In Table 10-2, it is apparent that the only real difference within each optical disc format is the recording technology. ROM means that these discs are Read-Only Memory—which means that the user cannot record to them at all. These are discs that you buy in the music or video store, with the music or video already recorded on them. They cannot be recorded or erased, which is why they are termed ROM.

Recordable discs (which include the types CD-R, CD+R, DVD-R, DVD+R, and BD-R), can be recorded by the consumer, but only one time. Rewritable discs (which include the types CD-RW, DVD-RW, and BD-RE) can be recorded and erased multiple times by the consumer. The basic operation of each of these recording technologies will be explained in the subsequent sections.

ROM RECORDING TECHNOLOGY

The end result of the ROM recording technology is to produce tiny bumps, as seen in Figure 10-15. When the focused read laser hits these bumps, some of the laser light is scattered. Where there are no bumps, the laser light is not scattered and most of it reflects back into the detector. Thus, where there is a bump, less light is reflected than where there is no bump. This simple difference in the amount of light

Format	Recording Technology	Approximate Capacity	1x Data Rate	Laser Wavelength	Track Pitch	Disc Size (Diameter, Thickness)
CD	ROM	750 MBytes	1.2 Mbps	780 nm	1600 nm	120 mm, 1.2 mm
CD	Recordable	750 MBytes	1.2 Mbps	780 nm	1600 nm	120 mm, 1.2 mm
CD	Rewritable	750 MBytes	1.2 Mbps	780 nm	1600 nm	120 mm, 1.2 mm
DVD	ROM	4.7 GBytes	11 Mbps	650 nm	740 nm	120 mm, 1.2 mm
DVD	Recordable	4.7 GBytes	11 Mbps	650 nm	740 nm	120 mm, 1.2 mm
DVD	Rewritable	4.7 GBytes	11 Mbps	650 nm	740 nm	120 mm, 1.2 mm
BD	ROM	27 GBytes	36 Mbps	405 nm	320 nm	120 mm, 1.2 mm
BD	Recordable	27 GBytes	36 Mbps	405 nm	320 nm	120 mm, 1.2 mm
BD	Rewritable	27 GBytes	36 Mbps	405 nm	320 nm	120 mm, 1.2 mm

Table 10-2: Optical disc formats and technologies and their associated characteristics.

reflected is interpreted by the optical disc drive as our digital ones and zeros.

These tiny bumps are produced by a plastic injection molding process. A master stamper is first made using an electron-beam process, which allows features of almost any size, including much smaller than these tiny bumps. The master stamper can be used for making many thousands of discs, each disc taking only a few seconds. After the plastic (usually polycarbonate) is injection-molded against this master stamper, it contains tiny pits where the bumps were on the master stamper. Then a very thin layer (about 30 nm) of reflective material (usually silver or aluminum) is deposited over the entire surface of the disc, including these tiny pits. Then a final protective layer is deposited over this reflective layer, and we thus have all the essential layers of a ROM optical disc: the bumps in the polycarbonate, a thin coating of a reflective material, and a layer to protect the reflective layer.

Discs made by the ROM recording technology are very inexpensive to produce, if made in large volumes (5,000 or more). The main costs are the master stamper, the plastic injection molding machine, the polycarbonate, the reflective material (although very little is needed for such a thin layer on each disc), and the protective material. Since

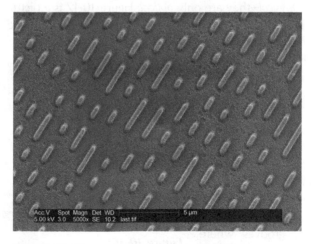

Figure 10-15:
CD-ROM bumps. The smallest bumps measure approximately 800 nm long and 800 nm wide.

the material costs are quite low, the other main costs of the master stamper and the plastic injection molding machine are amortized over the very large volumes of discs that can be produced by this method. A modern machine for manufacturing optical discs can generally run unattended for many hours, making tens of thousands of discs every day under essentially full automation. But it is entirely impractical to use this manufacturing process to make low volumes of discs.

ONE-TIME RECORDABLE TECHNOLOGY

Because these discs can only be recorded once, they are often referred to as "one-time recordables". Although they can be read in the same drives as the ROM discs, their recording operation is entirely different. Figure 10-16 shows a small portion of recorded CD-R disc. While the tracks are readily apparent, no data can be seen. Figure 10-16 is an SEM (scanning electron microscope) picture, which means it only sees things which are physical—bumps, pits, etc. An SEM cannot see things which are only optical, such as sunlight shining though a prism and onto a piece of paper. The SEM could see the paper, but not the rainbow of colors produced by the prism. Likewise, the SEM cannot see the data marks created by the recording process, because they not *physical*—they are only *optical*. Figure 10-17 is a picture of this same CD-R, but taken with an *optical* microscope; here the data is clearly visible on the recorded portion of the disc.

Like ROM optical discs, a recordable optical disc is also made primarily by plastic injection molding with a master stamper, but this stamper has only the tracks on it—there is no data on the stamper. So once the polycarbonate disc is stamped, it contains tiny grooves where the tracks are, as seen in Figure 10-16. The entire

Figure 10-16:
SEM picture of a CD-R. The track pitch is the same as for Figure 10-15.

surface of the disc is then coated with a light-sensitive dye; this dye particularly fills these small grooves. Then, as with ROM optical discs, the entire disc is coated with a reflective layer and a protective layer, and the main functional layers of the disc are complete.

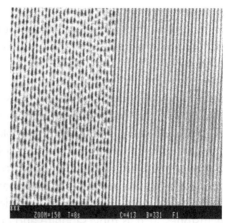

During the recording process, the reflectivity of the light-sensitive dye is permanently altered by a relatively high-power laser, as seen in the recorded section (the left side) of Figure 10-17,

Figure 10-17:
Optical microscope picture of the same CD-R shown in Figure 10-16.

creating our ones and zeros. When the lower-power read laser attempts to read this data, where the dye has been unaltered, it absorbs much of the focused laser light, as seen by the unrecorded section (right side) of Figure 10-17. Where the dye has been altered by the recording laser, much more of the laser light is reflected, as seen in the lighter portions of the left side of Figure 10-17. Although the contrast is not huge, it is sufficient to give us reliable data, and recordable optical discs are still a huge and healthy market.

REWRITABLE RECORDING TECHNOLOGY

Optical discs which are rewritable use a technology which is different still from the previous two recording technologies, although there are still several things in common. The layers are very similar to the four layers of the one-time recordable optical discs: the polycarbonate layer is injection-molded with the grooves which are the tracks; there is a layer of a special recording material; then there are the reflective layer and the protective layer, as with the recordable optical discs. So, the main difference is the special recording material, and it is indeed unique!

Molecular structures

In order to understand the unique property of the recording material used in rewritable optical discs, it is first necessary to understand something about the molecular structures of solid matter. This subject is also covered in Chapter 3. Most solid matter is usually in one of three molecular structures: amorphous, polycrystalline, or crystalline. These molecular structures have a great deal to do with the properties of the solid materials. A familiar example of this is carbon.

When carbon is pure and in its crystalline molecular structure, the atoms are all evenly spaced, with no irregularities (see Figure 10-18a). This is a highly stable form of carbon that we know as diamond. Other examples of crystalline materials include quartz (crystalline glass) and gemstones such as ruby, sapphire and emerald.

When carbon is pure and in its amorphous molecular structure, the atoms are irregularly spaced (see Figure 10-18c), the bonds are weak, and the material is soft and slippery. The most familiar example of this is graphite and soot. Other examples of amorphous materials include glass, plastics and most rocks.

The polycrystalline molecular structure is in between the crystalline structure and the amorphous structure. It is characterized by small regions ("grains") of crystalline orientation of molecules, separated by misalignment and irregularities ("grain boundaries"—see Figure 10-18b). This molecular structure occurs most commonly in metals, and the hardness of a given metal is dramatically modified by changing its molecular structure.

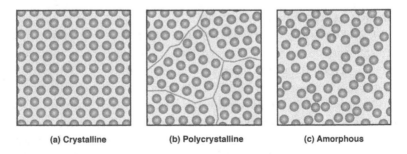

(a) Crystalline   (b) Polycrystalline   (c) Amorphous

Figure 10-18:
The three molecular structures in solid matter.

In some materials, their optical properties are also significantly affected by the molecular structure. This is very familiar from the example of carbon: the crystalline form is hard and optically brilliant when properly cut, while the amorphous form is soft and black and cannot be polished into optically brilliant crystals.

Chalcogenides

There is a class of materials known as the chalcogenides (meaning mainly the materials sulfur, selenium and tellurium in Group 16 of the Periodic Table) which are highly reflective in their crystalline state, and dull (much less reflective) in their amorphous state. While this is optically very interesting, it would not be worth much unless there is a reasonably easy way to get these materials to change back and forth between these two states. And since we're now focusing in on these materials, that must mean that there IS a way to do this.

If you heat a chalcogenide up to just *above* its melting point, the molecules arrange themselves somewhat randomly, which is amorphous. If we then cause that melted region to cool down very quickly, the molecules will remain in this amorphous state when solidified. This then becomes the dull, less reflective state, which we could call our digital zero.

If you heat a chalcogenide up to just *below* its melting point for a short period of time, the molecules arrange themselves into a highly ordered state in a process known as *annealing*. If we then cause that annealed region to cool down to room temperature, the molecules will remain in that crystalline state. This then becomes the highly reflective state, which we could call our digital one.

These thermal processes are diagrammed in Figure 10-19, with figure (a) showing the thermal profile for producing the amorphous molecular structure, and figure (b) showing the thermal profile for producing the crystalline molecular structure.

Because the chalcogenide materials used can be changed to either their amorphous structure or their crystalline structure, they are rewritable. While this rewriting process cannot be repeated an infinite number of times, it has been shown that it can be repeated many millions of times, which is sufficient for nearly all applications.

163

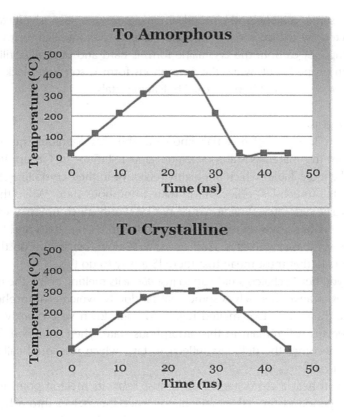

Figure 10-19:
Thermal profiles used for producing the amorphous and crystalline molecular structures in rewritable optical discs.

SUMMARY OF OPTICAL STORAGE

All of today's optical data storage is stored on CDs, DVDs, or BDs, all of which share a disc which is the exact same in outer dimensions. All of them come in ROM, recordable, or rewritable technologies. All of them are nonvolatile. And all of them are removable, meaning that the user can have a library of these discs (not so with HDDs). The discs are generally quite inexpensive, and the discs seem to remain usable for many years.

As with solid-state and magnetic storage, there are always the downsides. Some of the weaknesses of optical disc storage include its slow access time, its slow data transfer rate, and its relatively small storage capacity, as compared to HDDs.

The slow access time of optical storage is due to the significantly larger mass of an optical read/write head—dozens of times more massive than magnetic read/write heads. The laws of physics always impose a significant time penalty in trying to move a large mass quickly to a new location. The typical access time for optical storage is about 150 ms, or about 20 times slower than HDDs.

The slow data transfer rate of optical storage is due to the presence of a single read/write head; HDDs usually have at least four read/write heads. Additionally, the physical data density (number of bits per unit area) is less than on HDDs, so not as much data comes by the head in the same amount of time. In the end, the data rate of a 4x BD (generally about the fastest optical storage today) is 144 Mbps. While this is pretty good, it does not compare well to today's fastest HDDs, which come in at over 10 Gbps.

The smaller storage capacity of optical discs comes primarily from having much larger bits in optical storage than are presently possible with HDDs. A single bit on a BD (the highest density optical storage) measures about 300 x 300 nm; in the previous section on HDDs, we mentioned that a single bit on a HDD measures about 40 x 40 nm, which in terms of areal density means that HDDs are about 60 times more dense.

Error Detection and Correction in Computer Storage

In Chapter 8, there is a section on FEC (Forward Error Correction), particularly as applied to deep-space communications. This very powerful technology has also been successfully applied to data storage, and the benefits have been tremendous. Figure 10-20 is from the DVD consortium, and shows that, for a typical bit error rate of 1 bit in 200 (shown as a BER of 1E-2), the error rate *after* error detection and correction is performed goes up to 1 bit in 100 *sextillion* bits—an increase of 18 orders of magnitude, or 1,000,000,000,000,000,000 times! This is with an overhead of only about 25%.

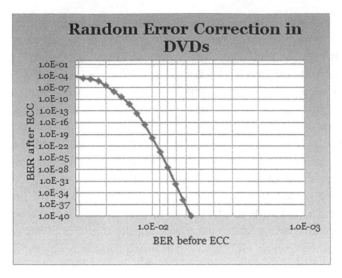

Figure 10-20:
Error detection & correction, as applied to DVDs. The improvement in error rate is about 18 orders of magnitude.

Yes, an overhead of 25% does seem like a lot, and it is, but with compression (covered in Chapter 6 in the section *"So How DO You Stuff an Elephant Through a Straw?"*) the overhead is not a problem, and the *huge* reduction in the error rate is well worth it. In fact, if it were not for ED&C as applied to data storage, today's incredible data densities would not be possible or even approachable, and the data you store on any of the preceding media would not be very reliable.

Chapter Take-Aways

Today's storage options for digital data are many, and overall they are quite affordable and reliable. Most readers of this book probably own a few of each of these types of storage (with the exception of magnetic tape), and some may have several. They each serve very well in their own respective niche. I hope that the next time you use one of these amazing storage technologies, you think of the incredible details happening at the physical and electronic level, many of them enabled by the integrated-circuit manufacturing processes discussed in

Chapter 4. And the size of the bits is something that will continually amaze me—hundreds of bits in a space smaller than the human hair—it's almost impossible to believe!

Despite the incredulity of what has happened already, future advances are to be expected in all these areas. New types of solid-state storage are nearly as dense as DRAM yet nonvolatile, and are only a few years away. Tape drives and hard-disc drives will continue to increase in density and remain in the same price range. And there are many formats for optical storage presently being researched; only time will tell which (if any) of these make it to a commercially successful product.

References/Additional Reading

1. Leavitt, David, "The Man Who Knew Too Much: Alan Turing and the Invention of the Computer", Atlas, 2006.
2. Nardo, Don, "Computers: Mechanical Minds (Encyclopedia of Discovery and Invention)", Lucent Books, 1990.

CHAPTER 11
CONVENIENCE OR CURSE: CELL PHONES

The cell phone has truly become ubiquitous in modern society, and has transformed most of our communications expectations. If we roll back the clock a few decades, we'll see a time when the only phone a person had was either at home or at work. There was only one phone at home, and it was fixed in its location in some (hopefully) central room in the house. If we were neither at home nor at work, we were generally without access to a phone except through use of a pay phone.

The cell phone has totally changed all that. We no longer are without access to a phone for any extended period of time. People usually expect to be able to reach us no matter where we are, and no matter the time of day—and we expect to do the same! Whether this ready availability is a convenience or a curse mostly depends on the individual and the situation, but anyone who has owned one cannot deny the reality of both!

First, a Tour of the Phone System

In order to understand how cell phones work, we must first understand the underlying technology that allows voice to travel over wire, over fiber, or just to travel through the air (wireless). This is fundamentally one of the oldest electronic technologies we have today, starting well over 100 years ago.

GETTING VOICE TO TRAVEL OVER WIRE

Sound travels in waves, much as light and radio waves, but in a completely different form. Sound of any kind must have air in which these waves exist and travel; a vacuum does not allow sound to propagate through it. And sound travels at a mere 750 miles per hour (1210 km/hr)—fast, but nowhere near as fast as the 670,000,000 mph (1,080,000,000 km/hr) that light and radio waves travel. So how do you make voice travel over wire?

Although doing this is quite simple now, it was certainly not that way to begin with. Today, all that is needed is a microphone—a device which converts the sound waves into electrical signals, which can then be transmitted over wire. But what would you do if there were no such thing as a microphone? How would you go about inventing it?

Every material has at least three fundamental electrical properties: resistance, inductance, and capacitance. These properties can be sensed and measured quite readily, so the basic challenge becomes one of making the waves of sound cause some kind of change in one of these electrical properties. Once we do this, we then have an electrical analog of sound. And once we have that electrical analog of sound, we can then transmit it like any electricity, and at the speed of electricity.

One of the first ways of creating this analog of sound was with a pack of carbon granules, as depicted in Figure 11-1. Carbon is somewhat conductive, so we can pass current through it relatively easily. And the amount of current that passes through this carbon granule pack is a function of how tightly the granules are packed together— tighter packing lowers the resistance. If we arrange a diaphragm such that the vibrations that strike it also cause the packing density of the carbon granules to change, we will have a resistance variation which is an analog of the sound producing the vibrations. If we then pass a current through this varying resistance, the current will produce that same analog. After a few other electronic refinements, this current is ready to be our voltage, which can be amplified and transmitted long distances over wire at the speed of light (or very near to it).

Over the years, many other types of microphones have been designed and improved, each one producing a small change in the resistance, inductance or capacitance (or a combination of these) of a material, which in turn is used to create a voltage which is an analog of the sound we wish to transmit. Today's microphones are so tiny and so effective that they are almost indistinguishable in a cell phone—they just do their job, and they do it very well!

GETTING VOICE TO TRAVEL LONG DISTANCES

Once we have a voltage analog of the sound we want to transmit, we're still not quite there, since we need this signal to travel many

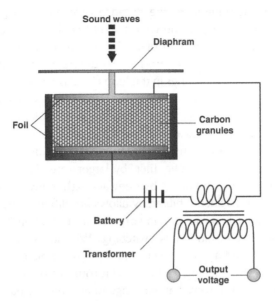

Sound waves

Diaphram

Foil

Carbon granules

Battery

Transformer

Output voltage

Figure 11-1:
The basic operation of a carbon-granule microphone, one of the earliest (and most rugged) types of microphone.

miles—sometimes thousands of miles. And there's the problem—the farther such a signal travels over wire, the weaker the signal gets. This is known as *attenuation*—simply the loss in amplitude of the original signal. It occurs primarily because the wire over which the signal travels is not without its opposition, known as resistance. Every foot or meter that the signal travels causes some of that signal to be lost, until eventually so little is left that we can no longer retrieve the original signal.

One main solution to this problem is to place amplifiers periodically along the line, to boost the signal. The magic of amplification is that it takes a very small signal and uses it to control a much larger signal, thus producing in the much larger signal a nearly exact copy of the original signal. Amplification can be compared to a person moving a small lever, which in turn causes a drawbridge to move. The movement of the drawbridge is proportional to the movement of the lever, but much larger in force and in distance.

Another main solution to the distance problem is to group telephones into small clusters known as a *local loop*. In rural areas, a local loop might be a few miles in diameter, and include 10 to 20 telephones. In metropolitan areas, the loop would be much smaller in size, and the number of telephones in the loop could be much larger. These telephones are all grouped together into one cable with several pairs of wire, which can handle several phone calls simultaneously. Then several of these local loops are grouped together into an *exchange*, and so on up in the hierarchy—much as road traffic is handled first by small streets with slow-speed traffic, then by larger multi-lane streets with higher traffic speeds, and finally by freeways with many lanes and very high traffic speeds. Such a hierarchy allows amplifiers to be placed in optimal locations, and allows many phone calls to be handled at once.

The other main solution is switching. Without switching, every phone would need a line to every other phone, which is completely impractical. With switching, a phone call is routed from one local loop to an exchange, and from that exchange to another exchange, and so on until the final exchange is reached, and then to the final local loop, and finally to the right telephone. Each of these switchpoints performs an essential function. Originally, this switching was done manually, by operators who plugged cords into the right place to connect the call to the next step in the route. Later, this switching function was performed by relays which were operated by the pulses put out by the dialing action of the phone dial. And now this switching action is performed by computers, making this switching one of the most automated functions in the world.

With amplifiers where needed, the hierarchy of local loops and exchanges, and appropriate switching, it is possible for a phone call to travel literally around the world. Earth-bound distances are not an issue, since the signals travel at near the speed of light. Even for a VERY long-distance call of 15,000 miles, the signal would only take a few milliseconds to reach the desired destination.

The preceding situation sufficed for many decades, and even overseas long-distance calls were usually very successful, with the conversant sounding like they could be in the same town. Customers appreciated this and came to expect this high quality in all their phone conversations. But there is always another element trying to impose major limitations on this success: noise. Noise has many sources, and while much can be done to reduce the impact of these noise sources, as long as the signal remains analog, it is highly susceptible to noise. And once the analog signal is corrupted by noise, it is extremely difficult to remove that noise.

Let's Go Digital!

It has long been known that if these analog signals could be converted to digital signals, noise problems would be minimized, if not eliminated. But digital circuits were more complicated, and for many decades after telephones became widely available in the USA, going digital was just not economically feasible.

All this began to change in the 1960s after the invention of the integrated circuit (see Chapter 4). Soon after integrated circuits were invented, digital circuits began to look practical, and soon the major links became digital, followed by the exchanges and finally even the local loops. While digital circuits do require a lot of additional complexity to take our analog voice signals and convert them to a digital representation, the advantages provided by going digital quickly outweigh the additional complexity.

The first advantage of digital which we will discuss is the dramatic reduction in noise. This is possible due to digital circuits called *repeaters*, which are basically digital amplifiers. As a digital signal moves along a wire, it loses amplitude and becomes corrupted by noise, just as an analog signal does. This is shown in Figure 11-2.

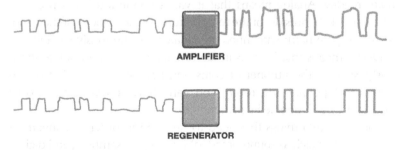

AMPLIFIER

REGENERATOR

Figure 11-2:
Digital amplifiers and regenerators can provide tremendous reduction in noise.

But one very big difference with digital signals is that there are only a few discrete voltage levels which are legitimate. A repeater has threshold circuits which can detect when a voltage is above or below these discrete levels, and the repeater then simply converts the incoming signal to a new, perfect digital signal, but *without* the noise. As long as this is done at appropriate intervals, the digital signals can be completely restored to their original, pristine condition, and the noise is almost completely eliminated, as also shown in Figure 11-2. This is truly one of the best advantages of going digital—this ability to dramatically reduce the noise.

One other advantage of going digital was discussed in Chapter 8—forward error correction (also known as ECC, for error correction coding). This advantage makes it possible to further tolerate noisy environments: conditions which would render analog signals incomprehensible are generally handled quite well with forward error correction.

So, by the 1970s, the transition to a digital phone system was well underway. While this transition would take more than just that decade, due to the vast installed base of analog phones, local loops, and exchanges, it was essentially completed before the turn of the millennium. It is true that, if you know what to look for, you can still find a few vestiges of the old analog system, but by and large, the USA and the world have gone fully digital for communication—and with no regrets!

CONVERSION FROM ANALOG TO DIGITAL

It is generally not obvious how one goes about converting an analog signal to a digital one, although we can understand the two domains quite readily. Analog means that it varies continuously, such as the amount of light outside, or the temperature, or the velocity of the wind. Most things in nature are inherently analog—continuously variable.

Digital means that it varies in discrete increments, and not continuously, such as the number of coins you have in your pocket or the number of people in a room: you can't have 4.376 coins in your pocket or 8.672 people in a room.

There are two things that need to be done in order to convert an analog signal (such as sound or video) into digital: sampling and digitizing. Sampling is just that—you take a quick sample. This is how film for movies works—the film does not have a continuous image of what's going on, but rather takes 24 pictures per second (called *frames*). These 24 samples per second, when played back through a projector, produce what our eyes and our mind interpret as continuous motion. The same is true for audio or voice—if we take the right number of samples each second, when we play it back we will have something that our ear and our mind will interpret as continuous sound.

Digitizing involves converting each sample to a digital number (a string of 1s and 0s). This can be easily visualized by thinking of a small stadium with 64 rows of seats. If we then fill the stadium with water and wish to digitally measure how deep the water is in the stadium, we could see which row the top of the water has reached. The answer would be a number from 1 to 64. We can then convert that number to a binary digital number from 000000 to 111111.

The process of analog to digital conversion is easily done with electronic circuits made just for this purpose. Just put in the analog signal and set it up right, and out of it you get a string of 1s and 0s representing the samples which have been converted to digital. The only remaining questions are: how often do you need to sample, and how many bits do you need for each sample? Research has shown that sufficient quality for voice (phone) is obtained with 8,000 samples per second, and 7 bits per sample. For audio (music), sufficient quality is obtained with 44,100 samples per second and 16 bits per sample—this is what you will find on CDs.

Then Let's Add Wireless Phones

So by the 1970s, the phone industry was in the process of going digital. And this was the same era when cell phones in the USA were being developed. The original cell phone system in the USA (now known as 1G for first generation) was analog, and it was the first cell phone system in the world. The technology necessary to make it a reality was amazing, but it was not cheap. Even in 1960s US dollars, the phones and their associated transmitter (which had to be mounted in the trunk) cost about $2,500, and a monthly subscription was about $100. It goes without saying that only the affluent could afford them, or those who truly had a need to stay in touch while traveling.

Second-generation (2G) cell phones were a major advance, and were digital. Their weight and size were dramatically reduced, and they no longer needed a separate transmitter. Their price was likewise substantially reduced, which quickly began opening up the market to those not quite so affluent. It was during this time that other countries began introducing their cell phones and installing cell phone systems in their countries.

Of necessity, cell phones operate without a wired connection—they transmit electromagnetic waves through the air. The challenges of transmitting data wirelessly are many (some are discussed in Chapter 7 and Chapter 8), especially when it comes to cell phones. Interference quickly becomes a very big issue: every other cell-phone user becomes a potential source of interference (noise) for your cell phone, and they must all be interoperable (meaning that the use of one cell phone cannot interfere with the use of another). They need to operate anywhere in the country. They need to be relatively secure, to make eavesdropping difficult. They need to be relatively reliable, so that customers will not be dissatisfied with the service. They need to be small (which also means a small battery), yet they need to last a reasonable number of hours between recharges. And they need to operate in a moving car, which is no small challenge!

And Now Add the Cells

One of the key technologies that has enabled today's cell phone was the concept of cells, for which the phones are named. Basically, there are multiple groups of frequencies on which cell phones can operate. There must exist some way to keep these frequencies from interfering with each other, and that is the concept of spatial division multiplexing—a complicated term which simply means we reuse these frequencies if they're separated by an appropriate amount of distance (space). While this sounds complicated, we're actually all familiar with this concept in another, closely-related area: radio and television stations.

For example, if I'm in Los Angeles and I tune to an FM radio station at 89.1 MHz on the dial, I will get a different radio station than the one I get if I tune to that same frequency in San Francisco. FM and AM radio stations all over the world reuse the same frequencies, if they are separated by enough distance that the radio stations don't interfere (or at least don't interfere very much). In a sense, we could say that Los Angeles is one cell, and San Francisco is another cell, and since they are separated by enough distance, these cells don't interfere. But what about San Luis Obispo (about half-way between San Francisco and Los Angeles)? Here one could get some of the signal from San Francisco and some of the signal from Los Angeles, and herein lies some of the challenge of spatial division multiplexing—at the boundaries of where these frequencies are reused, interference is inevitable.

This interference problem is solved for cell phones by assigning the different groups of frequencies such that where cell overlap occurs, the frequencies are different, as shown in Figure 11-3. This solution has reduced interference to a minimum, and has been widely adopted.

Figure 11-4 shows how the signal from cell towers (see Figure 11-5) spreads out and forms the cells depicted in Figure 11-3. Figure 11-4 also shows that a single cell tower generally provides for three cells, one on each of its three sides (see Figure 11-5).

The combination of cell towers and cell phones makes it possible for a cell to have a diameter of over 20 miles (32 km). Each cell is only capable of having approximately 2,000 subscribers. While this is adequate for rural cells, it is not sufficient for higher-density environments. But the number of subscribers per cell is fixed by the number of frequencies

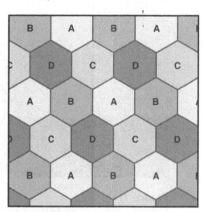

Figure 11-3:
Adjacent cells in the cell-phone system use different frequencies to avoid interference.

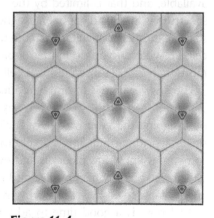

Figure 11-4:
Signals from cell towers grow weaker as you move away from the towers, represented by the small triangles where three cells meet.

Figure 11-5:
Examples of typical cell towers. The one on the right has been disguised as a ponderosa pine tree (mostly).

available, and that is limited by the spectrum available. The available spectrum is finite and has been assigned to many wireless applications, so it is extremely difficult to increase the number of subscribers in each cell. So what do you do in metropolitan areas where there are many more than 2,000 subscribers in a cell with a diameter of 20 miles?

The answer here is to simply reduce the size of each cell, by reducing the power of the cell towers, by building more cell towers, and by assigning the groups of frequencies accordingly. In some cases, a single apartment building is a cell; these geographically tiny cells are sometimes known as picocells and even femtocells. Special repeaters (electronic boxes which receive and retransmit the cell phone signals) are available to easily set up such tiny cells.

And now is a good time to address the question of how do cell phones work in subways and elevators? A subway is underground—so how do the signals from the cell tower reach the cell phone, and vice versa? And an elevator is surrounded by metal, so how do the signals from the cell tower find their way inside the elevator, and vice versa? The answer is that they don't—the cell tower signals (without some special provisions) DON'T travel down into a subway system, and they DON'T make their way inside an elevator. This problem is also solved with repeaters, and if you know what to look for, you can spot them in these settings. It is truly amazing that we have been able to make it possible for cell phones to work in these places, for it is not simple. But because cell phones have become so ubiquitous, we have become accustomed to having them work wherever we go.

Which brings up another little issue—most cell-phone users have experienced situations when their cell phone had no "bars"—no signal from a cell tower. This is either because their particular service provider did not have a contract in the city in which they were at the time (most cell towers are shared by multiple service providers), or because they were in a remote location where no cell service is available. Why would cell phone service providers not have cell phone service everywhere? It is simply a matter of economics. If there are enough subscribers in a given area, it is economically practical to provide service there. But if the area is remote (meaning there are very few subscribers in the area), it is simply not economically practical to erect cell towers and the necessary infrastructure to provide cell service to those areas.

And Finally: Add Endless Features

So, with the advent of the second generation of cell phones (2G phones), coverage was excellent, service was reasonably priced, and the cell phone system had gone digital. What else needed to be improved?

For a time, it seemed that future generations of cell phones would be unnecessary—other than to make them smaller. While the "necessity" of additional generations of cell phones is still arguable, the main answer to this question comes in one word: *features*. Cell phone manufacturers needed to find ways to distinguish their cell phone from their competitors' cell phones, and this generally was done through features and fashion. And the features that quickly became the most popular included texting, email, Internet access, GPS and maps, and downloads of every kind (especially including apps and ring tones). And why did these features require further improvements in the cell phone system? Well, some did, and some did not! More about that in the next sections.

TEXTING

Texting is unquestionably VERY popular, but its convenience can readily be argued. For example, in 2005, Jay Leno on the *Tonight Show* featured a competition between the USA's fastest text messager and some of the fastest Morse-code senders (a technology well over 140 years old)—and the Morse-code senders won! Clearly, text messaging is not the fastest nor the most convenient way to send a message; typing characters with two fingers (or thumbs) on such a tiny keyboard is clearly not as fast as using all 10 digits on our hands on a full-size computer keyboard. But its popularity has grown immensely, much to the delight of all cell-phone service providers.

The reason the service providers are so delighted with the popularity of text messaging is that it is almost free for them to provide, yet customers willingly pay for it! A bit of perspective is in order here.

To transmit voice, it takes approximately 56,000 bits of data per second, without compression. So, each cell phone has a single channel assigned to it, which channel must both transmit and receive 56,000 bits/second, simultaneously—an aggregate data rate of over 100,000 bits/second. Additionally, there is a small control channel assigned to each cell phone; this control channel only needs to send about 5,000 bits/second, and only one-way at a time—an aggregate data rate of only 5,000 bits/second. This control channel allows the cell-phone service provider to know which cell your phone is in, and to control the transmit power of your cell phone. But even at only 5,000 bits/second, this control channel is generally idle most of the time.

Sending a text message requires only hundreds of bits/second, so the unused time of the control channel works great for this. This control channel is something which the cell-phone service provider already has to have in place, so allocating a portion of the time of this control channel costs the cell-phone service provider essentially nothing—just the software to handle text messaging. So, providing text messaging costs the cell-phone service provider essentially nothing—but the revenues are fantastic! Therefore, the cost-benefit ratio is a no-brainer for the cell-phone service providers—they love it!

The bottom line on text messaging—there is no upgrade necessary to 2G cell phones. The basic cell-phone service is fully capable of handling dozens of text messages every hour.

EMAIL: IT'S THE ATTACHMENTS!

Originally, email was purely text-based, and (as mentioned in the preceding section), text messages do not require an upgrade on cell phones beyond 2G. However, over the years, people have become accustomed to attaching pictures and other files to their emails, and here's where it gets sticky.

A typical email is only a few hundred characters long and therefore only a few thousand bits long, so sending and receiving one via a 2G cell phone would only take a small portion of the cell phone's 56,000 bits/second voice channel. But a typical picture is easily 500,000 bytes in size, which is 4,000,000 bits! This would take over 70 seconds to send at 56,000 bits/second, and that's not even a very big picture! To say nothing of video files or high-definition pictures—these files are even bigger! Clearly, 2G phones were not designed for this application.

INTERNET ACCESS

The Internet is covered in Chapter 12, so we'll skip the details here. Suffice it to say that, together with email, these two applications (email and Internet access) have become the main drivers for 3G and now even 4G cell phones. Such phones provide the potential of data rates over 100 Mbits/ second, which allows rapid access to web pages, searches, downloading, and basically any type of Internet activity available through computers. All "smart" phones have this capability, and its popularity continues to grow. And to provide this kind of service, cell-phone service providers have had to upgrade their towers and cell-phone manufacturers have had to upgrade their cell phones—and thus the 3rd and 4th generations of cell phones have become very hot-selling products.

GPS AND MAPS

Map services, along with GPS, have also become extremely popular features on 3G and 4G cell phones. GPS is explained in detail in Chapter 15, so we'll skip the details in this chapter, but it is truly amazing that GPS receivers (originally many thousands of dollars) have become so small and so inexpensive that they can readily fit in a cell phone! And coupled with the ability to reference maps of nearly any place in the world, customers don't get lost as easily, and they can find places they might otherwise have never known about. What a great convenience!

Cell Phones and the Future

It is always fun to think about what the future will bring, especially to something that has impacted our lives as much as cell phones have. And my crystal ball is probably about as cloudy as anyone else's, making my predictions about as reliable as most prognostication. But sometimes the urge to futurize is just too great, so I'm going to venture a few guesses.

It is easy to look back and see what the cell phone has gone through and what the main drivers have been for these changes. Features have been the big sellers—this is not likely to change. So-called "smart" phones have truly shown the way here, with tens of thousands of "apps" being available for download, adding seemingly endless features to any "smart" phone. This is likely only to get bigger—more apps, more options, more features, even for the low-end phones.

The cost, size and weight of a cell phone have been fairly steady for several years now. The cost ranges from $40 to over $400, depending on features (and excluding specials available from the service providers); this is likely to stay the same. And while most customers wouldn't mind if their cell phone weighed a little less, they seem quite happy with the 4-ounce (130 g) weight common today. Size is not likely to change—an iPad is clearly too large, while a cell phone half the size of today's cell phones would be too easy to lose and would have a screen and buttons unacceptably tiny.

Will there be 5G cell phones? Undoubtedly. It is only a matter of time. There is plenty of impetus, for customers, for service providers, and for cell-phone manufacturers. And work on this has been underway for some time already.

Will they grow more complicated and easy to use? Certainly—yes to both! But how can they grow *both* more complicated *and* easier to use? Haven't we seen that already? Isn't one of the main attractions of Apple's iPhone how easy it is to use? And aren't nearly all "smart" phones more complicated to use?

Will cell phones become the eventual platform for video phone calls and video conferencing? Yes. So much of the infrastructure for this is already in place, and with adequate improvements in the coming years, (including built-in video projectors), cell phones will become the video and audio solution to staying connected with those we want to stay in touch with.

Chapter Take-Aways

The next time you make or take that all-important(?) cell-phone call, just think about the amazing complexity that underlies the working of that phone and the system of cell towers and phone connections. It is truly amazing that, in spite of the fact that any single point of failure is a serious problem, we still enjoy "5-9s" reliability (it works 99.999% of the time, or it only fails 1 hour out of 100,000). This is due to the built-in redundancy, which has been part of the phone system since its early beginnings, and which we have all come to expect (and hopefully appreciate!)

Think also of the way your voice is being sliced up into 8,000 samples per second, with 7 bits assigned to each sample, and that these 56,000 bits every second are flying through the air at the speed of light, to a (hopefully) nearby cell tower, which then handles the data stream and sends it on to the next necessary destination. And all this happens so quickly that you (usually) hardly notice the delay—that's really quite remarkable!

What is the future of cell-phone service? Worldwide, we now have more than 5 billion cell-phone subscribers, well over 2/3 of the world's population! And that's in spite of the fact that cell-phone service is terribly expensive in many underdeveloped countries. Clearly, cellphone service has become nearly ubiquitous, and it is likely to continue this trend. The future will provide more widespread service with fewer and fewer cell-phone "dead-spots". The number of dropped calls will diminish. The data rates of Internet access will continue to improve, beyond today's 3G and 4G to the 5G just off the horizon and other generations yet to be defined. Despite their tiny screens and keyboards, people will continue to use them to watch video clips, send, receive and look at pictures, browse the Internet, text message (there were more than 6 trilion text messages sent worldwide last year), and provide anywhere/anytime access to a world of information. It is extremely doubtful that this trend will diminish. In another decade, it is very likely that 9/10 of the world's population will have cell phones, despite the fact that some of them (children less than 3 years old?) would seem too young to join this throng.

So, whether we consider it a convenience or a curse, I would say the cell phone is definitely here to stay. I also believe that it will soon provide access to previously unthought-of resources that will bless (and challenge!) our modern society. Stand by for dramatic changes in personal health care, security, shopping, social networking, transportation, identification, and other areas presently untouched. It should get very exciting!

CHAPTER 12
HERE, THERE AND EVERYWHERE:
THE INTERNET

The Internet has truly become as integral a part of our society as automobiles, highways, electricity, and telephones. It would set our modern civilization back quickly and severely if the Internet were to suddenly disappear (which isn't any more likely than any other of these advances disappearing!)

Perhaps one of the first things we should do is clarify what is meant by "internet" and "the Internet", for these terms have grown to mean two different things. An "internet" is simply a network of computers which are connected to each other so that they can share data. As useful as this is, it pales in comparison to the usefulness of "the Internet", which means the worldwide interconnection of computers. The Internet interconnects everyone with access to it—users can send and receive information from any other Internet user, or from any of the hundreds of millions of websites that are out there.

A saying about the Internet is attributed to Robert Metcalfe: "The value of a telecommunications network is proportional to the square of the number of connected users of the system." While the exact quantification of the value of a telecommunications network is certainly subject to debate, no one can argue against the basic premise: the value of the Internet has increased as a function of the number of people connected to it. First begun in the 1960s, the number of computers connected to the Internet grew only slowly, and relatively few people in the world knew or cared about it. But with the introduction of a point-and-click interface (also known as a graphical user interface or GUI) in the early 1990s (also known as a web browser), coupled with the increase in home computing, its usefulness began to grow rapidly. And as it grew, its value increased greatly, so that in only a few years it became almost essential. Today, a computer isn't even considered fully a computer unless it has access to the Internet and the vast amount of information available through it.

First, a Tour of the Internet

In order to understand how the Internet works, we should first hearken back to another massive telecommunications network: the phone system (including cell phones), described in Chapter 11. Common elements are quickly apparent: the value of telephones grew as the number of phones grew, and when the majority of the people had phones, they quickly became a modern "necessity".

Another common element is the need for interconnections. Every phone needs to be able to connect to every other phone, through some kind of network of connections. In the phone system, this is done through a hierarchy of connections, also described in Chapter 11. With the Internet and fully digital switching, this has happened in a different way than it did for the phone system.

There is a hierarchy of connections with the Internet, as there is with the phone system, but the Internet hierarchy is much flatter and much wider. For example, the most typical way that your home computer is connected to the Internet is through a service provider—a company that sends and receives the data from your home computer's Internet connection (see Figure 12-1). This cable connects to your service provider's nearest group of servers (a server is simply a computer which is dedicated to communicating with other computers, usually many of them [hundreds or thousands]).

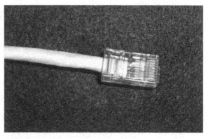

Figure 12-1:
An example of a computer's connection to the Internet. This is also known as an RJ-45 connector, an Ethernet connector, or a CAT-5 or CAT-6 cable.

Each time you attempt to access a web page, or download anything from a website, your home computer sends out a service request to your service provider. The servers at your service provider then check to see if the requested information is contained on those computers; if the requested information is there, these servers "serve" that information to the requesting computer. If the requested information is not there, these servers have an index to many of the millions of other servers which are also connected to the Internet, and this request for information is then sent to the servers where this information most likely resides. This forwarding of information requests continues until the requested information is located, and it is then returned to the requesting computer (yours).

Does this sound complicated? That's only because it IS. But because the hardware and software of servers has been developed over decades, and because millions of these information requests can be handled every second, the Internet is able to keep up with it, and handles the complexity with aplomb. But there is certainly more to the operation of the Internet than this.

THE ENVELOPE (PACKET), PLEASE

Communication over the Internet has another very significant difference from the phone system—it is NOT connection-based. If your computer needs to get information from another computer (usually a server), the Internet does not establish a connection as in the phone system. Instead, the data from the server is broken down into packets, with each packet containing some of the requested information.

It's much as if your only way of communicating with someone were with post cards. If you needed to send them a lot of information, you could number each post card and send the information via a sequence of these numbered post cards. You'd fill out each post card, and if it took you several hours or even days, you'd mail them out as you completed them. Then the postal service would find the best route for each of your post cards, and eventually all your post cards would reach their final destination. Each post card might travel a slightly different route, depending on what trucks or airplanes were available at the time. But once they all arrived at their destination, your receiver would then put them in the proper sequence (because you numbered them) and the information would be received correctly.

Let's take for an example that you're downloading a file. This file resides on some servers physically located in Kansas City, Missouri, and you're physically in Tallahassee, Florida (at least while you're attending a conference there). You send out the request for information. This is usually a message so short that it can easily fit on a standard Internet post card, called a "packet". This packet is received by your service provider, or by your hotel's service provider. The information is not on those servers, but their directory indicates it is probably on the servers in Kansas City, Missouri, so the packet is forwarded there.

The servers in Kansas City, Missouri, receive the information request packet so they check for the requested file and it's there, but it is so big, it will take 43 packets (each about 1,500 bytes long—sizes do vary), so the file is broken down into 43 sequentially-numbered packets and these packets are sent back to your computer in Tallahassee. But it gets a bit more complicated. On their way from Kansas City to Tallahassee, some of the packets encounter some traffic congestion. And just like we do when we see one street or highway becoming congested, we often try alternate routes. But in traffic, this only works well when we can pick our alternate routes intelligently—ideally when we have a chopper in the sky, telling us which routes are open.

This task of finding alternate routes on the Internet is done by *routers* (sometimes the names of these things actually make sense!) These highly sophisticated dedicated computers have many connections and are constantly monitoring traffic and sending out their little choppers in the sky (packets that test the connections), so they are always able to send incoming packets on to their destination (or at least one hop closer to their destination). And since distance is almost irrelevant, some of the packets may go first to Salina, Kansas, because the router may know that there is an unoccupied route from there to Pensacola, Florida. Once in Pensacola, it is only a hop away from Tallahassee, so the packet then hops from Pensacola to your computer.

So, eventually (over the course of many thousands of milliseconds), the packets of the requested file make their various ways, via many possible routes, to your computer in Tallahassee, Florida, at which point your computer puts the packets back together in their proper order and there you have it—the file you requested!

So now we've talked about servers, routers and packets, three very key parts of today's Internet. But there is certainly more than that! These pieces have been around for several decades, but the Internet has only become immensely popular in the last couple of decades. Some of the other key parts of today's Internet will be discussed in the following sections.

A Big GUI Mess—Or NOT

Since servers, routers and packets are not the key reasons for the proliferation of the Internet, let's talk about something that IS—graphical user interfaces, or GUIs. GUIs for the Internet are provided by software called *browsers*—programs such as Internet Explorer, Mozilla Firefox and Google Chrome. These browsers receive webpage information and display it in a way we (hopefully) find pleasing, informative, and useful, and they allow us to use the mouse or touchpad or touchscreen to select our choices.

Before there were browsers, webpages did not really exist as such. Though there were other ways of displaying information received from other computers, they were all text-based. Even highlighted text was unusual (underlined, **bold**, or *italics*); if highlights were desired, users would typically use techniques such as ALL CAPS or <<enclosing>> the text to be highlighted. Only one font and one text size was available. No graphics of any kind were used.

And we have to remember that, not too many years before Internet browsers appeared in the early 1990s, most computers didn't even have high-resolution color monitors—and some were still monochrome! And a few years before that, computers didn't have a mouse—just a keyboard! So, the elements that make our webpages so rich today (color, multiple fonts, multiple text sizes and emphases, graphics, and interactive graphics) were not even available.

So GUIs came at an extremely opportune time—it had only recently become feasible to provide nearly all of the preceding features, including ubiquitous mice, high-resolution color monitors, graphics, and multiple font and emphasis options; and Internet content was becoming more and more useful.

Browsers brought all of this together by adapting a technology which had been used in the printing industry for decades—markup languages. The first markup language for the Internet was the hyper-text markup language—also known by its acronym HTML—and this language still underlies much of the World-Wide Web (WWW). If you ever want to see what this stuff looks like, just choose a webpage you want to see, then right-click on an open portion of the webpage. One of the options you'll have is View page source. If you choose this option, you'll see the HTML that your browser interprets in order to make a webpage look like it does. If it looks complicated, that's only because it IS—but it sure works well! A well-written webpage is incredibly useful—information is truly right at your fingertips.

It's Somewhere in the Ether(net)

Another technology that has been key throughout the development of the Internet has been something known as Ethernet. This is simply a set of rules (a protocol) that have been agreed to by members of a standards body. This set of rules defined the signals, the media, and the information exchange procedure to allow computers to talk to each other. There were other protocols (such as token-ring), but Ethernet started out being the least expensive protocol that provided acceptable performance, and in the end, price won out. The higher-performance protocols (which were also more expensive) continued to lose ground, and today, the vast majority of computers connect to the Internet via the Ethernet protocol.

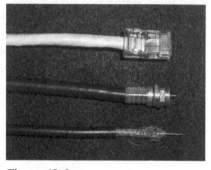

Figure 12-2:
Examples of media for connecting to the Internet. At the bottom is thin-wire coaxial cable; in the middle is thick-wire coaxial cable; at the top is the CAT-5 cable shown in Figure 12-1.

Robust protocols are very flexible, which allows them to adapt and grow as technology and applications change. Ethernet has truly been a star at flexibility. It can be used over multiple types of media (see Figure 12-2), although today twisted-pair wire (CAT-5 and CAT-6) is almost exclusively used. It can handle multiple data rates, including 10 Mbps, 100 Mbps, 1 Gbps, 10 Gbps, 40 Gbps, and even 100 Gbps!

Multiple computers can access the same Ethernet cable, as long as they don't do it at the same time. When they do access it at the same time, a data *collision* occurs, and the packets from both computers are generally lost. This is when a very interesting part of the Ethernet protocol takes over; this part is known as *carrier-sense multiple access with collision detection*, or CSMA/CD.

Any time a computer needs to access the Internet via Ethernet, it first checks to see if the connection is already busy. This is akin to a person wanting to back their car out into the street—the person first checks to see if the street is busy. If the street is devoid of traffic, the person backs out into the street (without looking as they back up). If traffic happened to come by while the person is backing up, a collision occurs. Then both drivers throw away their cars and go get new cars, then each driver waits a different random amount of time (a few milliseconds or less), then repeats the process. Usually collisions don't occur; but if a collision does occur, this random waiting period (different for each computer) generally assures that a second collision won't occur (at least between those two computers).

It is truly amazing to this author that such a strange protocol actually works! But because collisions don't actually destroy anything physical, and because creating new packets costs only time (and a few microwatts), it's actually quite practical and easy to implement, which is part of what has kept it inexpensive. And as long as the shared Ethernet connection doesn't get too busy, it works predictably and reliably.

Another thing should be mentioned about Ethernet connections. Since they are predominantly over twisted-pair wire (more about this in Chapter 13), their distance is quite limited—usually no more than 100 meters. While 100 meters is plenty for connecting to other nearby computers or perhaps to your service provider, it is by no means long enough for connecting to the servers which may be a few miles away. So how does the Internet connect between routers, when the distances can be many miles? The answer to that is usually through optical fiber, which is also described in Chapter 13.

Another technology, discussed in detail in Chapter 8, is cyclic-redundancy checking (CRC), which is used extensively in Internet communication. It is also known as the checksum, and is an integral part of most Internet packets. It allows the receiving computer to determine if any of the bits of the packet have been corrupted. Each time a packet is successfully received, the receiving computer sends a brief ACKnowledgment (ACK). If any bits have been corrupted, the CRC checksum in the packet will detect this corruption. When that occurs, the receiving computer sends a Not ACKnowledged (NACK), and the transmitting computer sends that packet again.

There are other communication details which take place over the Internet, using another protocol known as the Transmission Control Protocol/Internet Protocol, or TCP/IP. This protocol has also been extremely robust, governing most Internet information interchange for several decades.

But What If You Want a Continuous Flow?

The TCP/IP and CSMA/CD protocols were designed for communication where the data is not very time-sensitive. Examples of data which are not time-sensitive include email and downloading files—if the packets are briefly interrupted or take a few seconds longer than normal to be delivered, it is not a problem. These protocols have worked very well for these applications.

But what if you're using your favorite IP telephone program (such as Skype), or listening to a radio station, or watching television or a movie on the Internet? These activities have become increasingly popular over the Internet, and have created some unique challenges. In these cases, if some packets are delayed too long, the flow is seriously interrupted and stalls occur; if too many of these stalls occur, this form of Internet entertainment loses its value very quickly. These are examples of what are known as *real-time* communication, meaning that the flow of packets is very time-sensitive. At issue here are performance parameters known as *latency, jitter, packet-loss ratio,* and *throughput rate.* A bit about each of these is important here.

Latency is simply the amount of time it takes a packet to go from the transmitting computer to the receiving computer, and is primarily a function of the number of routers the packet has to go through (distance is almost insignificant, unless we get into geosynchronous satellite links). And latency is really only important for 2-way communications such as phone calls—but when it reaches values greater than about 200 ms, you'll really notice the difference, and phone calls become awkward as you begin talking when the other person was already talking. Anyone who has experienced this is very familiar with how challenging such phone calls can be.

Jitter is simply the variation in the latency. If some packets arrive very quickly (a few milliseconds), while other packets take many milliseconds to arrive (remember that they can all take different routes), the jitter is high, and the receiving computer has to wait a long time to buffer all the incoming data so it can put the packets back together in the right order. Jitter is only a problem if you're doing real-time communication, where the flow needs to be continuous.

Packet-loss ratio (PLR) is simply the percentage of packets that don't make it to their destination, for whatever reasons. If the PLR rises above about 2%, the number of packets that are being retransmitted, plus the gaps in the flow, result in significant interruptions. These interruptions are familiar to all cell phone users when their coverage is getting poor, and they usually tell the other person they're "breaking up". We hear lots of gaps (lost packets) in the flow of the sound, and conversation quickly becomes very unpleasant.

Throughput rate is simply how many bits per second can reliably be expected over a given Internet route. IP phone services require the least throughput rate. Video requires a higher throughput rate than audio, and HD video requires even more throughput than standard-definition video.

Again, these performance parameters are not a major issue, unless we are dealing with real-time communication, where a continuous flow of data is essential. It is truly amazing that TCP/IP and CSMA/CD, which were NOT designed for real-time communication, have been so successfully adapted to these applications.

Is Cyberspace Safe?

Cyberspace is a term coined many decades ago to refer to a computer-based existence. Many of us spend a great deal of time in cyberspace, accessing social media, stores, games, and many other websites. Like any place where we can spend our time, there are always risks that unwanted intrusions can happen. These intrusions have given rise to a host of cyberspace risks, including computer viruses, spyware, adware, identity theft, spoofing, Trojan horses, etc. Given the risks, it is important for all cyberspace visitors to take appropriate precautions, as we would when traveling anywhere.

Some of the simplest and yet most successful ways to counter these intrusions include passwords and security software, in addition to making sure you know well anyone who has access to your computer.

PASSWORDS

Experience and research have shown that a good password will stop most computer intrusions, and certainly deter the casual attempt. A good password is one that cannot be easily guessed by an outsider, and also cannot be easily guessed by a computer. The two approaches are very different.

Most of us use passwords which are easy to remember, because they have personal meaning to us. Examples include the name of our spouse, child, dog, or favorite teacher. This means that, if an intruder is someone who knows us well, they could easily guess our password.

One way a computer attempts to guess a password is by using a dictionary—any words in that dictionary can be quickly tried, and computers being used for intrusion on other computers can guess millions of passwords in a very short time. Another way a computer attempts to guess a password is to try all the simple, default (unchanged) passwords, since there are only a few dozen of these.

So the best rules on passwords: 1) Don't use a name someone else can easily guess if they know you; 2) Don't use words in your password; 3) Do use at least one number and one special character; 4) Do use at least 8 characters; 5) Do make sure you change the default password on any computer equipment you purchase.

Unfortunately, it is inevitably true that the best passwords are also the most difficult to remember. Nevertheless, if we wish to keep our online accounts and our computer safe, it is very important to use passwords which are difficult to guess. Ahh, the agony and the ecstasy of complexity!

SECURITY SOFTWARE

It is not the purpose of this book to recommend any particular product, so I will avoid naming specific programs or services. Suffice it to say that if you wish to keep your computer safe from outside intrusion, some security software is essential. Features you should look for include: 1) Anti-malware, virus, Trojan, spyware, worm, rootkit, and phishing; 2) Browser exploits; 3) Secure network; 4) Inbound and outbound email protection; 5) other more advanced features, depending on your specific needs. And after you install the software on your computer, be sure to set the password(s) as per the rules above!

KNOW YOUR USERS

Anyone you give access to your computer also has ready access to everything on it. Social engineering takes advantage of this, wherein a would-be intruder uses their social skills to persuade you to give them access to your computer, either in person or remotely. Be sure such would-be users are people you know and trust; violations of this type are some of the worst.

The Future of the Internet

Ahh, this section is one where we can truly dream big. In the late 1990s, during what became known as the dot-com boom, it seemed that everyone would put something up on the Internet (particularly on the world-wide web) and that brick-and-mortar businesses would soon fade away. The subsequent dot-com crash later showed these expectations to be overblown. But the fact remains that the Internet, particularly the world-wide web, has changed the developed world in huge ways, and this trend is unlikely to change. Some of the world's largest companies did not even exist before 2000. Some of the services we take for granted have only been available for less than a decade.

So, based on this quick backward glance, it seems inevitable that major changes will continue to occur, and that the future will bring services to our home that we do not even envision presently. Basic Internet performance parameters, including latency, jitter, PLR and throughput rate, will continue to improve. This will allow these and other new services to exist.

Looking back, when electricity first became widely available, people could not possibly have imagined all the future technologies it would enable, including television, cell phones, the Internet, and everything else discussed in this book and much, much more. The widespread availability of the Internet will certainly do this as well.

A small example is in order. Surgeons have demonstrated the ability to remotely operate a robot, which is used to perform surgery on a patient thousands of miles away. This was done over the Internet. Such an application requires a reliable, real-time communication link, and making such a link affordable, even commonplace, has been a major triumph of Internet technology. It would never have been attempted only 20 years ago. Given another 20 years, surgeons could have a virtual presence, being able to see, feel, hear, and be aware of all that is happening to a patient anywhere on the planet, and to respond accordingly. Gamers will have virtual-presence games that will be hard to separate from reality—and that is not necessarily good, but it seems inevitable that it will happen, and the Internet will probably be the communication backbone for this.

I see both exciting possibilities and concerning possibilities; certainly we have both with us already. White-collar crime is at an all-time high. Child pornography is more widespread and readily available than ever before. Yet the opportunities that the Internet brings are also wonderful. We can only hope that the vast majority of all people using the Internet will use it for the betterment of mankind, or it could be the downfall of modern civilization.

Chapter Take-Aways

In my opinion, the Internet is the most significant thing to happen to computers since their invention. It has fundamentally changed much of our social and business world, and many other aspects of our lives. It has also created risks heretofore unknown. It is as much a part of modern society as electricity, the automobile, air transportation, running hot and cold water, and public transportation. It will continue to grow in importance in the decades to come.

The world today consists of many billions of computers with access to the Internet, including embedded computers. This number will continue to grow. Planning for increased connectivity is essential, as it will certainly be a major part of the future.

And the next time you do anything on your computer (desktop, laptop, netbook, tablet, smart phone, etc.) to interact with the Internet, just think of the incredible technologies that are at work, enabling the whole system to operate as effectively as it does. It is an incredibly complex system, made up of hundreds of sextillions of transistors, billions of nodes, millions of servers, hundreds of millions of routers, and trillions of packets flying about every year. It is truly a marvel that such complexity can all work as well as it does!

CHAPTER 13
TO INFINITY AND BEYOND!—
OPTICAL FIBER

There are basically only three media over which we send digital data: copper wire (including twisted pair and coaxial cable), wireless (meaning through the air or through the vacuum of space), and optical fiber. Chapter 12 briefly discussed sending data over copper wire, for access to the Internet; Chapter 8 discussed sending data through the air, or particularly through the vacuum of space (which is essentially the same, as far as radio waves go). Clearly, each of these media has its own advantages, or they would not all be used as much as they are.

The cheapest media would clearly have to be no media at all—wireless. While this is true for the media, it is not true for the system as a whole. A complete communication system has to include the transmitter, the receiver, and the media. But if we compare apples to apples, a wireless communication system, even though the media is free, is still more expensive than a comparable performance wired system.

Copper wire is still less expensive than optical fiber, with twisted pair wire being significantly less expensive than coaxial cable. The main use of twisted pair wire is for Ethernet connections and for the phone system; the main use of coaxial cable is for television.

Figure 13.1 shows a small section of the above media, along with a small section of optical fiber. At its heart, optical fiber

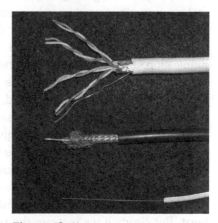

Figure 13-1:
Small lengths of: 1) twisted pair or CAT-5 cable (top); 2) coaxial cable; and 3) optical fiber (bottom).

197

is just pure glass, but to make it into a practical cable, there are many other things added, but more about that later.

Why We Like Fiber in Our Telecommunications Diet

So we've already established that, of the three main types of media available, optical fiber is the most expensive. So why is it so popular these days? Its advantages are several, and I think the best way to present them is in order of their significance.

AND THE WINNER IS: CAPACITY!

One of my favorite quotes about this comes from Frigo, Iannone & Reichmann, from the August 2004 issue of *IEEE Optical Communications*, p. S16: "... experts seem to agree that FTTH [fiber-to-the-home] is the most capable infrastructure for delivering *future* broadband services to homes. In fact, many believe that once fiber is installed, it will become the *only* wireline access medium since it offers *essentially unlimited bandwidth* and thus greater service opportunities than copper and wireless media." (emphases added). Basically they are saying that, once fiber is installed, you can continue to add capacity to it forever, for its bandwidth is essentially unlimited. But this confuses two terms: *bandwidth* and *capacity*. Although they are not the same, they are closely related, and are sometimes used interchangeably.

Bandwidth is described in Chapter 6, and is basically the size of the pipe (pumping water is often used as a comparison for data communication, and it works most of the time)—the bigger the pipe, the more water you can put through the pipe in a given amount of time. *Capacity* is a little more complicated, and has to do with the overall throughput of the water pumping system. It is a measure of how much water can be pumped through a given pipe, given all the relevant constraints. For instance, if we double the pressure of a water pumping system, we can almost double the flow rate without changing the size of the pipe. But there are always some limits to this—we cannot double the pressure as many times as we wish, for eventually the pipe would burst. But clearly if we increase the bandwidth (the size of the pipe), we also increase the capacity.

A mathematician named Claude Shannon derived a formula many decades ago for computing the capacity of a given telecommunications pipe (connection), as a function of the size of the pipe (the bandwidth) and the other relevant constraints. This formula, not surprisingly, is known as Shannon's Law, and has been shown to be true for several decades. If we apply Shannon's Law to optical fiber communication, we come up with a capacity of approximately 6.5 Ebps (E stands for Exa, which is 10^{18}, or 10 followed by 18 zeros. To put this into perspective, this capacity would be sufficient to provide a dedicated Internet connection of about 1 Gbps for every person on this planet! Another way to look at this is to realize that this is enough capacity to transmit every radio station in the world, every television station in the world, every phone call in the world, and every Internet communication in the world, all simultaneously, over the same single piece of optical fiber, and still have enough capacity left over to do the same for 9 more Earths just like this one!

Clearly, we do not yet have the technology to allow us to use all this capacity, but some impressive feats have been demonstrated. A recent demonstration transmitted 78 Tbps over a distance of 100 km (T = Tera, or 10^{12}, or 10 followed by 12 zeros). While this is indeed impressive, it is only a small portion of the capacity of optical fiber. Suffice it to say that once optical fiber is installed, all you need to do to increase the throughput is to install new transmitters and receivers—the capacity can always be increased (if you can afford the new transmitter and receiver!)

AND MISS RUNNER-UP IS: EMI IMMUNITY

Electro-magnetic interference (EMI) is forever with us, and is constantly evident with wireless and copper wire data transmissions. Its causes are many, and it is readily noticed when you're listening to an AM radio station during a lightning storm—every time there is a lightning strike (a very powerful source of EMI), you can hear the static it produces on the AM radio station. Copper wires pick up EMI quite readily, and any wireless system is particularly vulnerable to this, since the receiver in a wireless system is actually looking for electromagnetic signals that contain data, and EMI is just another form of an electromagnetic signal—which means it is just interfering noise.

But optical fiber is not susceptible to EMI at all! No matter how much EMI there is around a length of optical fiber, as long as we have properly modulated the light passing through the optical fiber, the EMI does not affect our optical signal at all. While part of this advantage is obvious (no one likes noise in their signal!), another part of this advantage goes right back to the winning advantage discussed in the preceding section: it actually improves the capacity, as compared to wireless and copper wire systems. That's because noise is one of those constraints that Claude Shannon identified as a limiting factor in the capacity.

If optical fiber were susceptible to EMI, that susceptibility would increase the amount of noise in the optical signal. And Shannon's Law tells us that if the amount of noise increases, the capacity goes DOWN. Conversely, since EMI does *not* increase the amount of noise in an optical signal, it also does *not* decrease the capacity of the optical fiber as it does the capacity of copper wire or wireless systems.

And there is another advantage this characteristic brings. Optical fiber signals do not *cause* EMI either. What this means is best understood by another example. Fluorescent lights often produce a significant amount of EMI; this can be experienced by placing an AM radio near a fluorescent light and tuning in to a weak station (or no station at all)—the noise created by the EMI of the fluorescent light is clearly audible. Electrical wires of all kinds, if they have electricity running through them, produce EMI. So if you put a lot of electrical wires close to each other, the wires with **large** amounts of electricity will produce EMI which will interfere with the signals on the wires with very **small** amounts of electricity—much like the big bully interfering with the little kids on the playground. But you can bundle optical fibers together as much as you wish, and the optical fibers with **large** signals will *not* affect at all the optical fibers with **small** signals. This is a very nice advantage!

VERY HIGH SECURITY

A secure signal is one that only goes where you want it to go, is not decipherable by anyone except the person for whom it is intended, and cannot be duplicated without someone detecting that duplication. Wireless signals are very poor in this first category, for they go just about everywhere. But even signals over copper wire are not considered secure by this measure, as it is easy to sense the signals in a copper wire without even touching the wire. This is because of the small amount of EMI being produced by the signal in the wire—it can be picked up just by having a detector close to the wire. And this type of eavesdropping is very hard to detect unless you can actually see where someone has placed their eavesdropping device.

But optical fiber is very different here. As we mentioned in the previous section, signals in optical fiber do not produce EMI, so placing an eavesdropping device near an optical fiber is ineffectual. The only way to eavesdrop on a signal in an optical fiber is to actually tap into the fiber, stealing off some (or all) of the photons—and this is easily detected at the receiving end, since some (or all) of the photons that normally would be arriving at the receiver suddenly aren't there anymore. This is referred to as *tamper evident*, and optical fiber is very tamper evident.

This is not to say that eavesdropping on optical fiber is impossible; I am aware of instances where this has successfully been done. But it is extraordinarily difficult, and this high level of difficulty greatly increases the security of signals that are transmitted through optical fiber.

Lowest attenuation

Attenuation is simply a loss of signal strength, and its causes are many. While much work goes into reducing attenuation, it is always true that in any communication system, not all of the signal that is transmitted reaches the receiver. In fact, usually very little of the original signal reaches the receiver—from 1/1000th to 1/10,000,000,000th, or even much less! This attenuation is a major limiting factor in the capacity of the communication system (again, part of Shannon's Law). Of the three media, wireless generally has the greatest attenuation, followed by copper wire. This means that, of the three media options, optical fiber has the lowest attenuation—by far.

This was not always the case. In the early days of optical fiber (in about the 1960s), it had a terrible amount of attenuation, even worse than most copper wire and some wireless systems. But optical fiber manufacturers have, over the decades, improved the manufacturing process to the degree that today's optical fiber loses less than 1% of what copper wire loses. A record set not too many years ago (2007) was a data rate of 2.5 Gbps over 7,500 km of optical fiber, *without an amplifier or repeater!* This record was a great example of the very low attenuation characteristic of optical fiber.

Another way of measuring and comparing the attenuation of optical fiber versus copper wire or wireless is to specify the amount of signal that is lost as a function of the distance. For instance, an ideal media would lose 0% of the signal in one kilometer—but no media is ideal. Wireless communication systems commonly experience losses of 99.9% (30 dB) in one kilometer; wired communication systems commonly experience losses of 99% (20 dB) in one kilometer. By contrast, optical communication systems typically experience losses of only about 7% (0.3 dB) in one kilometer.

ASSORTMENT OF OTHERS ADVANTAGES

Optical fiber has a few other advantages, but it is much harder to rank them, so let's just present them. It is lighter than copper, which has major advantages in airplanes and satellites, where weight is a critical factor. It is smaller than copper, which is also advantageous in airplanes and satellites, where volume is also a critical factor. And finally, it cannot accidentally cause a spark, which is an advantage in explosive environments such as oil refineries or grain elevators.

How Light Stays Inside the Fiber

Since optical fiber is made of extremely high-purity glass, it is fairly easy to understand how light can travel a long distance in the glass. But there's a little problem with a fact that is familiar to all of us—light travels in a straight line. Just turn on a flashlight in the dark and it's easy to see that light doesn't bend around corners. So how do you get light to stay inside the optical fiber when the fiber is not perfectly straight? For example, Figure 13-2 shows a child's toy which is simply plastic optical fiber connected to a flashlight. Notice how the light seems to bend with the plastic optical fiber, even doing a 180° turn! This is no optical illusion—the light is actually doing just that—bending. Yet there are no mirrors involved.

The answer to how light does this in optical fiber is a property of light called refraction, the evidence of which is seen quite readily. For example, Figure 13-3 shows a spoon in a glass of water. It would appear that the spoon has been seriously bent or broken where it enters the water, but we know that is not the case. We just accept it as a matter of fact that a spoon in a glass of water looks weird for some reason. Other examples of refraction of light are looking at the corner of an aquarium and seeing the same fish in two very different apparent locations; or the sparkling of a well-cut diamond; or a rainbow near a rainstorm. All of these things appear this way because light does not travel at the same speed in all materials, and this is what causes light to refract (bend).

One of the fundamental constants of classical physics is the speed of light in a vacuum (in space, for example), which is about 186,000 miles/sec (300,000 km/sec). And light passes through many transparent materials, including water, diamond, plastic and glass. But because these materials are much different from a vacuum, or even from each other, light travels at a different velocity in each of these materials. Table 13-1 summarizes this for some familiar materials, giving the *index of refraction* and the velocity of light in these materials.

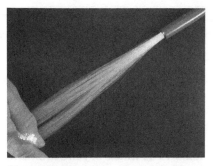

Figure 13-2:
Light from this small flashlight goes out straight ahead, but readily negotiates a 90° turn to come out the end of the fibers.

Figure 13-3:
It is the refraction (bending) of light through the glass and the water that causes this spoon to appear seriously bent where it enters the water.

	Space (vacuum)	Air	Water	Glass	Plastic	Diamond
Index of Refraction	1.0000	1.0003	1.33	1.52	1.58	2.417
Velocity of light (km/ sec)	300,000	299,910	225,563	197,368	189,873	124,120

Table 13-1: Index of refraction and velocity of light for several familiar materials. Note that light travels much more slowly in materials with a high index of refraction.

IT'S THE DIFFERENCE THAT COUNTS

While this property of refraction is intriguing, it is insufficient to answer the question that is the title of this section. And if all materials had the same index of refraction, (even if it were as high as diamond), we'd never notice it. In fact, we only notice refraction when there are materials with a *different* index of refraction, and these materials are directly in contact with each other. In the examples given above, we notice the spoon in the glass of water because the glass and water are surrounded by air, and air has a very different index of refraction from glass and water. A diamond is also surrounded by air, and the big difference between the index of refraction of air, compared to that of diamond, causes the brilliant colors seen in a well-cut diamond. An aquarium is surrounded by air, and again the difference in the index of refraction between the water and the air causes the same fish to appear to be in two different places.

Any time there are two materials with a different index of refraction in contact with each other, this interface causes light to bend, as though it were hitting a strange kind of mirror. And the severity of the bend is proportional to the difference in the indices of refraction. This is well demonstrated in Figure 13-4, which shows how a common glass beaker can be made to effectively disappear. The first photo shows the individual beakers, which are then placed inside each other in the second photo. The final photo shows how the smaller beaker has been made to "disappear", even though it is still inside the larger beaker. The larger beaker is readily visible because it is surrounded by air, which has a much different index of refraction. But in the third photo, the smaller beaker has been surrounded both inside and outside by a fluid which has the same index of refraction as the glass. This means that light is neither bent by the smaller beaker, nor absorbed by it (since it is transparent glass), thereby making it invisible.

So, going back to the example of Figure 13-2, the light from the flashlight enters the front end of each of the plastic fibers and attempts to travel in a straight line. But eventually some of the light hits the outside edge of the plastic fiber, which is surrounded by air. This interface is characterized by a great difference in the indices of refraction (1.58 for the plastic, 1.0003 for the air), which causes the light to bend—and it bends *inward*. This keeps the light inside the fiber, as though the fiber were surrounded with a mirror. And this happens with almost all of the light—very little of it ever escapes out the sides, so almost all of it makes it to the other end of the fiber, even when the fiber is bent!

Figure 13-4:
The property of refraction of light is what causes the inside beaker to apparently disappear.

NOW FOR THE REAL THING

The example of the plastic optical fiber is fairly easy to follow, which makes it useful for understanding. But actual optical fiber used for modern communication is not plastic, and it is much smaller in diameter, and it does not depend on being surrounded by air for it to keep the light inside the fiber. So, actual optical fiber is a bit more complicated, but not too much more. Figure 13-5 shows the preforms from which actual optical fiber is drawn or pulled, and Figure 13-6

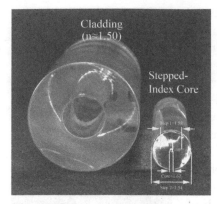

Figure 13-5:
The indices of refraction for a preform of optical fiber.

depicts two steps in the actual manufacturing of the glass fiber. There are several kinds of optical fiber; the kind depicted in Figures 13-5 and 13-6 has a core (the part where the light actually travels) with a relatively high index of refraction. This core is surrounded by a layer of glass with an index of refraction which is slightly lower, followed by another layer of glass with an index of refraction which is lower still. Each successive outward layer has a lower index of refraction than any of the preceding inward layers. Finally comes the cladding, which has the lowest index of refraction. These layers are consolidated together into a *preform*—a rod of glass about 2½ inches (5.25 cm) in diameter and about 1 meter long. This rod is heated up in a carefully controlled, fully automated machine and hot, almost liquid fiber is pulled from this preform at a steady rate until the entire preform has been made into thousands of meters of fiber.

The fiber thus produced has the exact same cross-section as the preform. It is the layers of glass, each with a successively lower index of refraction, which keeps the light inside the fiber even when the fiber is not straight. And it is the extremely high purity of the

Figure 13-6:
Preforms of optical fiber. On the left is the core of the preform, through which the light actually travels. In the middle is the hollow cladding, whose lower index of refraction keeps the light inside the core. At the right a the consolidated (finished) preform from which some fiber has been drawn or pulled; it is about 2½" (5.25 cm) in diameter at its largest.

glass that gives it the very low attenuation.

Figure 13-7 shows a section of optical fiber *cable* (cable is a carefully arranged group of multiple fibers, strength members, and other members) which is rated for burying; the

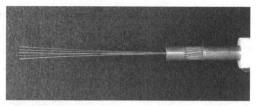

Figure 13-7:

A small section of optical cable with 5 fibers. This cable is about 1" (2.5 cm) in diameter at its largest.

vast majority of the cable shown is for strength purposes (to keep the glass fiber from breaking or being bent too much). At its very center are 5 optical fibers, each not much larger than a human hair. Figure 13-8 shows a section of optical fiber which can be hung or buried; again, the majority of the cable is for strength and protection of the fiber. At the core of this section of optical fiber cable are 5 bundles of 12 optical fibers; again each fiber is not much larger than a human hair.

What's with All That Wasted Space?

The first section of this chapter indicated that the best thing about optical fiber is its essentially unlimited bandwidth. Yet Figures 13-7 and 13-8 show cables with multiple optical fibers. If each fiber has an essentially unlimited bandwidth, why do we need so many fibers?

The answer lies in the flashlights (actually lasers) on the transmitting end, and the detectors on the receiving end. To use all of the bandwidth of an optical fiber, we would need lasers that could be switched on and off at an essentially infinite rate—billions of trillions of times per second. While today's lasers are fast, switching at over 100 billion

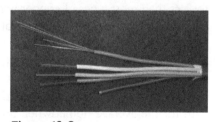

Figure 13-8:

A small section of optical cable with 5 bundles of 12 fibers each, for a total of 60 fibers. This cable is about ½-inch (1.5 cm) in diameter at its largest.

times per second, this is nowhere near fast enough to fully utilize all the bandwidth of a single fiber. And even if we had a laser that could switch billions of trillions of times per second, we do not have detectors that could sense those incredibly fast pulses of light, nor the electronics fast enough to process them if we could detect them.

And there's another limitation, which has to do with the number of different "colors" (usually called wavelengths) of light we can send down an optical fiber. A single optical fiber can have essentially an unlimited number of different wavelengths occupying it—at least in theory. The practicality of it quickly gets in the way, however. It rapidly becomes impractical and prohibitively expensive to put together the equipment necessary to put thousands of different wavelengths down a single optical fiber. Today's limit of practicality is about 500 different wavelengths; in a few years this could be 5,000 or more. But to put tens of thousands of different wavelengths down a single optical fiber will probably take us many more years. And yet even then, the fiber will not be full!

Chapter Take-Aways

There is no doubt that optical fiber is here to stay, at least for the foreseeable future. Future optical fiber links will have faster lasers, faster detectors, more wavelengths per fiber, and lower costs. The attenuation of optical fiber will continue to decrease, though not nearly as much as in the early years. The cost of optical fiber will also continue to decrease, and eventually a way will probably be found to produce it for less than most types of copper wire communication media.

The above improvements in optical fiber mean that the applications for it will continue to increase. Fiber to the home (FTTH) is becoming much more practical; eventually one of the several variations of this technology will displace copper to the home. But will fiber soon be installed inside your home? I don't think so—it's too difficult to work with for the average homeowner, and the advantages it provides are not really necessary once the signal is inside your home. Today's FTTH systems bring the fiber literally into the home, but once there the signal is converted to the standard twisted-pair and coaxial cable connections that our computer and our media equipment expect, and there isn't much need for this to change.

Some futurists envision 100 Gbps or even 1 Tbps connections to every home. This is certainly possible, since it can be done today but it is prohibitively expensive. However, there is ample evidence that this cost will continue to decline, and you may one day have that massive flow of data into and out of your residence! Imagine being able to download an HD-resolution video of a 2-hour movie in only a few seconds! Since this is already being done in some specific settings, it is almost undoubtedly only a matter of time (and the continual onward march of Moore's Law and of improvements in manufacturing optical fiber) before it becomes widespread.

CHAPTER 14
CONSUMER MANIA—
AUDIO AND VIDEO PRODUCTS

The history of audio and video products is full of outstanding successes—and a few flops. But it doesn't take much observation to see that our thirst for these products has continued unabated—and indeed seems to have grown—over the past several decades. Although the first products, by today's standards, would be considered low fidelity (the sound or the pictures were not exact replicas of the original) and very expensive, they were still quite popular and became fairly widely adopted.

Analog versions of these products included Thomas Edison's wax cylinder phonograph; records; reel-to-reel tapes; cassette tapes; 8-track tapes; radio; television; film cameras and video cameras; camcorders; and other related products. While most of these products are no longer popular (with the notable exception of radio and television), billions of people spent billions of dollars on these products in their lifetimes, and these products were generally very successful.

And the only reason most of these products are no longer popular is simply that they have been superseded by better ones. Reel-to-reel tapes, cassette tapes, 8-track tapes and records have all been replaced by the CD (compact disc); film cameras and video cameras have been replaced by digital cameras and video cameras. An entirely new class of audio/video product has been introduced: mp3 players and portable video players. And finally, video games in all their many forms have created an entirely new industry which is larger than the movie industry in terms of sales. Each of these deserves a look.

Digital Audio, Video & HD Video: CDs, DVDs & Blu-ray

The technology behind optical discs (CDs, DVDs and Blu-ray discs) is discussed in detail in Chapter 10. The advantages of digital audio, video and HD video are many, with the foremost being no noise in the background, very high fidelity, and a very wide dynamic range. Let's make sure we know what these terms mean.

211

Cassette tapes, 8-track tapes, reel-to-reel tapes, and records all suffered from a significant amount of background noise, which usually grew worse as the tapes or records aged. Once the noise in these media increased, it could not be decreased, and it was notoriously difficult to filter out this noise from the desired music or speaking. It was particularly noticeable during the quieter passages of the music or the pauses in the speaking, and quickly became rather annoying. But digital audio, video and HD video brought with them the inherent low-noise advantage of digital; even in the very quiet passages of music or the lulls in speaking, it is rare to be able to hear background noise, unless there is a problem in the system somewhere.

High fidelity is another characteristic that has become well identified with digital media. Fidelity simply means that the reproduced signal sounds very much like the original signal in all ways. The main things that usually reduce fidelity include noise (see previous paragraph), limited dynamic range (see next paragraph), and nonlinearities. Nonlinearities are imperfections in the recording process in which the recorded signal does not completely resemble the original signal due to limitations of the media or the recording process itself. Figure 14-1 shows what a nonlinearity would look like in an audio signal; what it would *sound* like is a lack of fidelity—something about it would not sound like it should.

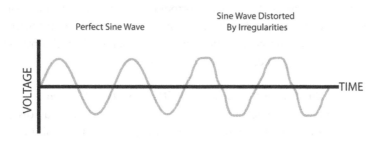

Figure 14-1:
An example of a perfect signal (first half of the wave on the left side) and a signal distorted by noise or by nonlinearities.

Dynamic range is simply the difference between the loudest sound and the quietest sound. In normal life, we commonly encounter a dynamic range of over 100 dB (that means a dynamic range of 10,000,000,000 in sound energy!) For example, if you're sitting in your home in the dark of the night with no heater or air conditioner going, no fans on, no traffic outside, and everyone asleep (and not snoring!), the few other distant or muffled sounds you'd hear would be around 10 to 20 dB. Then if you later crank up your car stereo to your favorite song, it can easily reach 110 dB, which is a dynamic range of 90 to 100 dB. In music, we don't usually need 100 dB of dynamic range, even in classical music, but we tend to prefer as much as possible. Professionals tell us that 90 dB of dynamic range is usually enough. But records were generally capable of only about 60 dB in dynamic range, and cassette tapes only about 45 dB. This was mainly because of the background noise that was always there, as discussed above. This is where CDs shine: they have a very silent quiet part, and a very powerful loud part, for a dynamic range of about 90 dB. And the same is true for DVDs and Blu-ray discs—they have a terrific dynamic range.

WORKING WITH BITS

Music, video, and HD video are inherently analog signals. But CDs, DVDs, and Blu-ray discs are inherently digital. So somewhere in between recording the desired music, video or HD video, these signals must be converted to digital. This conversion process is described in more detail in Chapter 11 in the section "Conversion from analog to digital". The conversion numbers for each of these three signals are fascinating, so let's take a look at them.

We'll start with audio, since it is the simplest of these three signals. Audio (meaning music, not just voice as for phones) consists primarily of signals in the range from 20 Hz to 20,000 Hz. Frequencies outside this range are generally not considered audible to human ears. In order to produce a high-fidelity digital representation of these frequencies, we need to sample it at least 40,000 times each second, with a resolution of 16 bits per sample. Standards have set the actual sample rate to 44,100 samples per second. This means that every second our analog-to-digital converter is producing:

(44,100 samples/second) * 16 bits/sample = 705,600 bits/second.

That's a lot of data! And it means that to store 75 minutes of music on a CD, we would need:

705,600 bits/second * 75 minutes = 3.175 Gbits = 397 Mbytes.

And that's just for one channel! Remember that modern music is stereo, meaning that it contains two separate signals, one for the left ear and one for the right. So, altogether, storing 75 minutes of digital music requires about 794 Mbytes! That is no small amount of data, especially for the mid-1980s when CDs came out (remember that hard-disk drives in those days rarely had more than 10 Mbytes).

It gets even better for video. DVDs were invented to store video at the resolution of normal television (see Chapter 6 for more about television). In that chapter, it is mentioned that normal television (sometimes called NTSC television) requires 525 horizontal lines, with 384 pixels on each line, for a grand total of 201,600 pixels (of each color). If each pixel is represented by 8 bits, we get:

3 colors * 201,600 pixels * 8 bits/pixel ≈ 4.84 Mbits.

And since television requires motion, we have to have enough frames per second to make it look like continuous motion. This frame rate for television is set at 30 frames/second. So this means that a digital signal for NTSC television would require:

4.84 Mbits/frame * 30 frames/second ≈ 145 Mbits/second ≈ 18 Mbytes/second.

And that's a lot of data! But to store a 2-hour movie on a DVD, we would need:

18 Mbytes/second * 2 hours ≈ *130 Gbytes!*

Well, 130 Gbytes is *much* more than a disc that size can hold (their capacity is only 4.7 Gbytes), and putting more data on that size of a disc in the mid-1990s was not possible. So, the solution was the digital magic of compression (see "So How DO You Stuff an Elephant Through a Straw?" in Chapter 6).

And finally, let's look at HD video. HDTV gives us wonderful pictures with amazing detail—but at a price! And I'm not referring just to the cost of the HDTV and the Blu-ray drives for playing HDTV videos. The price I'm referring to here is the native bandwidth (or size of the pipe) required to broadcast these images. A typical HDTV has a resolution of 1,920 pixels on each of 1,080 lines; with 8 bits of resolution on each of the three primary colors, we get:

3 colors * 1,920 * 1,080 * 8 bits/pixel ≈ 49.8 Mbits.

And at a frame rate of 60 frames per second, native HDTV would require:

49.8 Mbits/frame * 60 frames/second ≈ 2.98 Gbits/sec ≈ 373 Mbytes/second.

And to store a 2-hour movie at HD resolution, we would need:

373 Mbytes/second * 2 hours ≈ 2.69 Tbytes!

Fortunately, video gives us many ways in which the signal can be compressed (again, see Chapter 6), which in the end reduces this storage requirement to only about 25 Gbytes (the capacity of a Blu-ray disc).

Is it worth it? Certainly tens of thousands of consumers think so, as they continue to flock to their favorite electronics retailers and drop their money for their new HDTV. But you should be the judge; and it is for sure that your vote will be cast with your pocketbook. Figure 14-2 shows magnified sections of standard television (NTSC), compared to a similar magnified section of HDTV. The difference is not hard to appreciate!

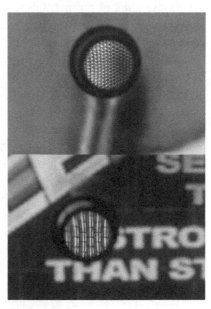

Figure 14-2:
The pixels of an HDTV (upper half), compared to the pixels of an NTSC TV (lower half). The HDTV pixels are noticeably smaller.

WHERE WILL IT ALL END?

The shift to HDTV begs the question of whether there is more yet to come. The ultimate answer to this question lies mainly in what the human eye can perceive. The world in which we live consists of essentially infinite resolution—things can get very small! But if an image contains pixels which are much smaller than what the human eye can perceive, what does it matter?

So, let's do a little experiment. Find a friend (or yourself) with fine hair (blondes tend to have finer hair than brunettes or blacks; babies have the finest hair; or a balding man); take one of their finer hairs, and then compare it to a hair from one of your thick-haired friends (or yourself). These hairs range in diameter from about 0.75 thousandths of an inch (about 20 microns) for the fine hairs to about 2.5 thousandths of an inch (about 65 microns) for the thick hairs. And if you look closely, without magnification (assuming your eyesight is 20/20), you can see the difference in the thickness of these hairs. This means that the human eye, from a distance of about 2 feet (61 cm), can resolve down to about 1 thousandth of an inch (25 microns).

So what is the size of a pixel on today's HDTV? Let's take an example of a 37" HDTV with resolution of 1,920 x 1,080. A 37" screen means it has sides of about 32" by 18", which means it has:

1,920 pixels/32 inches = 60 pixels/inch.

But the fact that our eye can resolve down to about 1 thousandth of an inch means that our eye can resolve 1,000 pixels/inch. That means there's a long way to go, at least as far as what we can see on the HDTV from 2 feet away. But let's be fair—nobody watches TV from 2 feet away (I hope!). Let's assume the average viewing distance to be about 10 feet (or 3 meters). This is five times farther away than the example given for a viewing distance of two feet. Our resolving power goes down approximately as the inverse square of the distance, which means that at 10 feet, our ability to resolve is only down to about 2.2 thousandths of an inch, or about 450 pixels/inch. There's still a lot of room for improvement!

But if SHDTV (super-high-definition TV, or about 500 pixels/inch) were to become a reality, it's going to be a long way down the road. Look how long it took to go from NTSC to HDTV—about 60 years! But there's another problem here, and that is the perceived improvement. Some people don't see a huge improvement between NTSC and HDTV, or at least not enough to justify the price right away. How much perceived improvement would there be between SHDTV and

HDTV? In my opinion (and at this point, it's mostly subjective), not much—and certainly not enough to justify the additional cost!

Let's try another example. This book was written on a laptop computer with a screen which measures about 13" by 8" (33 cm by 20 cm), and has a native resolution of 1,680 x 1,050, which means it has about 130 pixels/inch. That resolution is *twice* as good as HDTV, but does the average consumer see it as a huge difference? Generally, they do not—especially on moving images. Since movement tends to blur pixels together, super-high resolution on television is not nearly as valuable as on our computers, where the images are often static.

Bottom line: this author does not believe there is adequate margin for perceived improvement between our current 60 pixels/inch HDTV resolution, and the perceivable ultimate SHDTV resolution of about 500 pixels/inch. But if SHDTV ever does come about, it will probably be many years away, and it will not be adopted very quickly, simply because the perceived improvement is marginal.

So, there's my prediction for digital video. But what about digital audio? We've already established that today's CD features a sampling rate of 44,100 samples/second at a resolution of 16 bits/sample, for a data rate of 88.2 kbytes/second. This resolution seems to have sufficed for about two decades. In 2004, a new specification was released known as the High Definition Audio Specification, which can accommodate a sampling rate up to 192,000 samples/second, at a resolution of up to 32 bits per sample. While this new specification has caught on in limited circles, it has not caught the general public's fancy, and HD audio products are still very hard to find. In my opinion, it's for the same reason as prediction of SHDTV—minimal perceived improvement. It's very hard for most people to perceive any difference between the music from a CD (which is truly excellent) from that of HD audio. And because there is very little perceived improvement, the HD audio specification has not really caught on, and probably will never reach the interest level of the general public.

MP3 and Portable Video Players

Because compression is such a key part of making digital audio and video successful commercially, the Moving Picture Experts Group (a large group of experts in the fields of audio and video) released a standard in 1991 known as MPEG-1, which included standards for lossy compression of both audio and video. The part that specifically addressed audio compression (MPEG-1 Audio Layer 3) became known as MP3.

MP3 compression greatly reduced the size of audio files, an average of about 11 times. For downloading over the Internet, this truly became an enabler, creating the industry around which were built Apple's iTunes and other MP3 download sites such as eMusic, Napster, Rhapsody, Iomoio, and others. It also made possible the entirely new line of products we know as MP3 players, the most famous of which is Apple's iPod and its many versions. These players are amazingly small, light, affordable, and versatile, making them some of the most successful consumer products in the past decade.

Not long after the introduction of MP3 players, versions were introduced which included a small full-color screen. Some earlier versions had included a small monochrome screen, mainly intended for aiding the owner to choose the songs they wished to listen to. But these new full-color screens immediately created a new type of product: a portable video player (or digital photo viewer) which did not require a DVD! And although the screens are quite small, using them for these purposes has become very popular.

The root technology which made this all possible is compression, and specifically lossy compression. This is covered in more detail in Chapter 6, in the section "So How DO You Stuff an Elephant Through a Straw?" Suffice it to say here that most MP3 files provide digital audio at a fidelity which is sufficient for most consumers, yet at a file size which is much smaller than the original. This makes it possible to have a CD with over 150 songs, compared to the usual 14. An MP3 player with storage of 8 Gbytes is advertised to be able to hold about 1,000 songs. This is not enough for your entire song collection, but it is dozens of hours of listening pleasure. But if compression were not used, even 8 Gbytes would only hold about 90 songs—only a few hours, and only a very small part of your entire song collection.

Audio MP3 compression is actually very complex, and works in ways that remove the parts of the audio that the human ear is least able to perceive. It was developed over several years and was tested many times. It has been enormously successful, as judged by the wide public acceptance of this format and the many products based on it.

Video Games

When the first video games became available in the early 1970s, it was easy for most people to see that this would some day be a huge market. And as prescient as that perspective certainly was, it pales in comparison to reality. Annual sales in the USA are over $13 billion, which is bigger than the USA box office revenues for the movie industry. And this for an industry which did not even exist only 40 years ago!

The reality is that video games are extremely popular, especially among the younger generations. It is difficult to find someone under 40 years old who does not have a few favorite video games which they love to indulge in a few times a week. And because they are popular, it is fascinating to take a look inside these to understand a bit about their complexity.

Images on a computer screen must be either a reproduction of real things (your digital photos or movies), or they must be created artificially through what is commonly called *real-time computer graphics*. This term simply means that all of what you see on the screen of a computer in a video game is created through graphics calculations made at least 30 times per second (usually 60 times per second or more). Rarely is any of what you see during a computer game actually taken from real photos or movies—it is all generated artificially. And it is an incredibly complex process to make these images look realistic!

STARTING SIMPLE

Let's start with one of the simplest and earliest computer games, Pong. As seen in Figure 14-3, this game attempted to simulate a game of ping pong, with the television screen showing a very simple, squarish image of the ping pong ball, two simple lines to represent the two players' paddles, the black background to represent the table, and a dotted line to represent the net. The score was displayed right on the table. It permitted only very limited spin, and no lofting, slap shots, or anything else but simply meeting the ball when it came in your direction. For its time, it was advanced and entertaining enough to catch on fairly quickly. All the computer had to do was calculate the trajectory

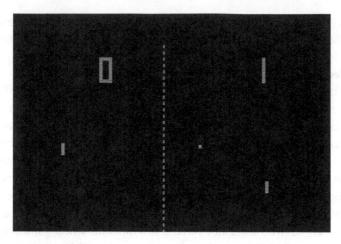

Figure 14-3:
A screen shot from the very early computer video game Pong.

and velocity of the ball and plot it on the screen, and calculate if the ball's trajectory carried it past one of the paddles or off the table. It also kept track of the score of each player.

In order to make this game entertaining, it had to be similar enough to ping pong so as to be engaging. It had to obey basic physics (ping pong balls don't suddenly accelerate, decelerate, or change directions), and it had to look something like what it was intended to represent. And it had only three moving objects—the ball and the two paddles, and the paddles could only move up and down. So the computation required by the computer was relatively minimal.

Now imagine a ping pong computer game with today's computers and today's video game expectations. It would need to allow spin, lofting, slap shots, moving the paddle anywhere, a full 3-dimensional representation of the table and playing area (including the room in which the table resides), the texture of the paddles, ball and all other surfaces, realistic rotation and trajectory for the ball, and so on. Although the setting is still relatively simple (only 2 players and one ball), the calculation requirements have increased dramatically. The players must have their movement appear realistic, in all positions of their hands, arms, legs, face, hair, etc. Shadows must be calculated in accordance with the lights in the room. Perspective must be adjustable to allow the players to see the game from any angle they wish. To do all this, the computer must keep track of every pixel on the entire screen, what color and intensity it needs to be, and what is happening to each moving object and its relation to everything else on the screen. If the player runs into the table, players would expect a realistic collision, perhaps including a realistic reaction from the player who ran into the table. And the physics of making such a collision look realistic are not simple.

Now multiply this complexity by several times as you enter today's multi-player environments where the scenery is constantly changing, the shooting must look somewhat realistic; and when players' avatars are hit, they need to fall in a realistic fashion. All these actions are happening to virtual objects—there is nothing truly real about them—yet the players must have the illusion of reality. Large objects take more energy to accelerate or stop than small objects. Wind can cause something to fall in a different direction. And all of this is done by calculations in the computer. The computer must go through subroutines dedicated to calculating the physics of the objects based on their simulated properties of mass, velocity, direction, etc. There must be subroutines for rendering the proper color, shading and texture of every surface; and subroutines for determining when one object should appear to interact with another object; and subroutines for defining how each object *can* interact with all other objects—if at all. The difficulty of doing this, especially in real time, is phenomenal, and is handled by special microprocessors known as GPUs, for graphical processing units. Many (if not most) of these GPUs are so powerful that they eclipse the performance of a computer's main microprocessor.

So, it should not surprise readers that modern video games take many people many months to write the programs for them, and that these programs are many millions of instructions long.

Bonus features

Video game programmers are like the rest of us in several ways, one of which is their enjoyment of having a little extracurricular fun. This has resulted in their programming in special features that have become known as *easter eggs*—because if you hunt a lot, you can find them. Just enter "easter eggs video games" into your favorite search engine and you'll find tons of websites that describe them, link to them, give the top 10, or whatever you're looking for.

To create an easter egg, a video game programmer simply writes a bit of "special" code (special meaning it's not essential to the operation of the game). This code causes something very unusual to happen, something which is not an integral part of the game. Surprise features like this are not unique to video games—they can be found in application programs, DVD movies, CDs, and other media. Go ahead and look up a few and give them a try—you'll enjoy it!

Chapter Take-Aways

So, a summary of this chapter would have to say that audio and video electronics have grown immensely in the past several decades, as measured by the number of products available, the amount of money we spend on them, the amount of time we spend with them each week, or almost any other measure. As always, there are both good and bad things to be said about this, but regardless of how we feel about them, they seem quite destined to be here for a long time.

As to what we might expect to see in the future, a lot of that has been addressed in each of the preceding sections. But about video games, some prognostication is in order. I believe we will continue to see increasing realism in video games (the current generation is truly amazing!). We will see new genres that haven't even been thought of yet, because of the increasing realism. Their popularity will continue to grow, along with the problems of addiction associated with that popularity. The good ones will become truly amazing and the classics will be with us for a long time. Nostalgia will capture many participants, as it already has in many genres.

There is one other thing that should be noted. The line between video games and video simulations is a very fine line; sometimes they are indistinguishable, as in flight simulators. I expect this will continue, and one field will feed the other—advances will be shared and will benefit both fields. This has been and continues to be a very fruitful cross-sharing of great development effort.

Some of the realism of video games and simulations has begun to bleed over into remote health care. This is, in my opinion, one of the best things that could possibly happen with this technology. Remote surgeries have already been performed, many times, and successfully. The same is true of remote diagnostics and treatment. I expect rather dramatic improvements in this area in the decades to come, and can't wait to see how much these improvements could mean to those people traditionally unable to access health care.

CHAPTER 15
YOU'LL NEVER BE LOST AGAIN—GPS

For many centuries before the modern era of electronics, navigation on the open seas was one of the thorniest problems in all of transportation and commerce. The magnetic compass was in use for navigation by the 11th century CE, and it helped greatly, but it could only tell direction—it could not tell you where you were on this sphere we call home.

For navigational purposes, the Earth is divided up into parallel horizontal lines (latitudes) and parallel vertical lines (longitudes). All measurement systems have to have some kind of reference. For latitude, the reference is the equator, which makes it the 0° reference point. Latitudes are called out in degrees southern (below the equator) or northern (above the equator) latitude. For instance, the latitude of New York City is at about 40° northern latitude, while Cape Town, South Africa lies at about 34° southern latitude, and the south pole lies (by definition) at 90° southern latitude.

For longitude, the reference is much more arbitrary: 0° longitude is defined as the line passing through Greenwich, England. All longitudes are measured in degrees east or degrees west of this vertical line. Using the same examples, New York City lies at about 74° western longitude, while Cape Town lies at about 18° eastern longitude. Halfway around the world from Greenwich, England, is the 180° latitude, which is also the International Date Line (with some notable excursions), and which latitude passes through Fiji and just east of New Zealand. The longitude of the north and south poles is not defined, for those are the points where all the longitudes converge.

The fixing of latitude has been solved for many centuries using tools such as the quadrant, the astrolabe, and the sextant, provided the skies were clear long enough to take a reading using either the sun or the stars. While this was still very difficult (a steady hand and a calm deck were almost essential), good navigators could fix their latitude quite accurately, many centuries ago.

The reason for this lies in the celestial clues available, and their relation to the Earth. If you are on the equator and the date is either March 20 or 21, or Sept 22 or 23, the sun will rise due east, be directly overhead at midday, and will set due west. There will be exactly 12 hours of daylight and 12 hours of nighttime. If you are north or south of the equator on these dates, this changes somewhat, and in a very predictable way. And if the date is other than one of these dates, it also changes the position of the sun in very predictable ways. And if the sun is not available during a cloudy day, the stars can be used to provide the same information. How far you are south or north of the equator can be directly observed based on the celestial bodies that are fixed with respect to the Earth, *if* you know the date.

But not so with longitude. Where you are with respect to 0° longitude can only be fixed if you know what your *local* time is, with respect to the time in Greenwich, England. While it is not too difficult to determine local time (the sun reaches its maximum altitude at noon, no matter your latitude), knowing the time in Greenwich requires a clock set to that time, and that clock must be accurate to within a few seconds each month. Such clocks were simply not available until the early 1800s, and even they were very expensive. Until each ship could have its own highly-accurate clock on board (known as the ship's chronometer), the problem of determining longitude was not solved, and it created great navigation challenges.

The magnitude of these challenges of navigation on the open seas, without modern technology, is hard to fathom. Probably one of the best stories about this is told in the book *Longitude*, by Dava Sobel (Walker & Co., 1995), on pages 17-20. A brief recounting of this story would be very helpful.

COMMODORE ANSON AND THE *CENTURION*

In September 1740, the *H.M.S. Centurion* set sail for the South Pacific from England under the command of Commodore George Anson. They had no chronometer. By March 1741, already 6 months at sea, the *Centurion* rounded the tip of Cape Horn, only to encounter a terrific storm that punished them for 58 days. When finally able to get a fix on their position, Anson sailed north and west for Juan Fernández Island for some desperately needed fresh water and food. When he reached the proper latitude, he had no idea which way to go—east or west. For over two months they had been unable to plot their progress, and they did not have a chronometer to allow them to determine their longitude. So on a hunch, he sailed west. But after not sighting land for four days, he decided he was already west of Juan Fernández, so they turned around and headed east.

Two days later, they sighted land, but it turned out to be the coast of Chile, under Spanish rule; they could not go there! So they executed another quick 180° turn and sailed west again. Finally, on June 9, 1741, they dropped anchor at Juan Fernández. The extra two weeks of sailing had cost Anson an additional 80 lives among his sick and ailing crew.

AND OTHER DANGERS

This story clearly shows one of the major risks of navigating prior to the invention of a good (meaning accurate and affordable) chronometer. But there was also another risk, which was even more dangerous and not very subtle. While some charts of the open ocean did exist, showing particularly the dangerous areas to avoid, the usefulness of these charts depended heavily on knowing where you were, and the open ocean offers no clues. There are countless stories of ships that wrecked themselves upon *charted* shoals, only because they did not know where they were and could not avoid the rocks in time.

When you really think about the magnitude of the problem of navigation on the open seas, it brings a great feeling of respect and admiration for the courage these early sailors had, not to mention their ability to navigate using such limited equipment. It also helps us realize just how big the need is for good navigational equipment.

LORAN: Electronic Solution #1

Early attempts to meet the need for something to accurately determine longitude were frequently related to consistent periodical events in the observable sky, including phases of the moon, the positions of the planets, and other changes in celestial bodies. These events were sufficient to give accurate calendars, so they could know what day it was, but they were insufficient for determining longitude.

Soon after radio became commercially practical, it was looked to for a solution to this problem of longitude. Many proposals were made, and the most widely accepted and successful of these was LORAN (LOng RAnge Navigation). Most of the U.S. portion of this system was completed shortly after World War II. This system consists of many land-based transmitters (usually near the coast), transmitting timing signals out over the ocean. The range was about 1,200 miles (1,930 km), which still left a large part of the ocean without electronic navigation assistance, but the most critical regions near the coastlines were effectively serviced.

The basic operation of LORAN is actually quite similar to how our ears locate the source of a sound. One of the clues our ears use is that of a time difference between when a sound arrives at our ears. If it arrives at our left ear before it arrives at our right ear, we know the sound is to the left of us. However, LORAN uses two transmitters and one receiver, rather than the two receivers (our ears) and one transmitter (the sound source) used by our ear-location system.

If two transmitters transmit consistent signals at a fixed difference with respect to each other, and if my ship is exactly the same distance from both transmitters, the fixed difference of the signal from the transmitters (usually a fixed number of milliseconds) will be the same. For example, if the transmitters send out signals which are exactly 2.50000 ms apart, and my ship is exactly 100 miles from each transmitter, the LORAN receiver on my ship will detect the synchronized signals from these two transmitters to be separated by exactly 2.50000 ms.

However, if my ship is closer to the first transmitter than the second, the LORAN receiver will detect the signals to be *more* than 2.50000 ms apart. Conversely, if my ship is closer to the second transmitter, the LORAN receiver will detect a time difference *less* than 2.50000 ms. And the magnitude of the time difference is directly proportional to the distance. If three different LORAN transmitters can be detected by the LORAN receiver, the position of my ship can be fixed within a few hundred meters.

LORAN has been a great success in navigational assistance. But even LORAN needed a very accurate timing reference. This need was met by atomic clocks which were periodically adjusted to align with

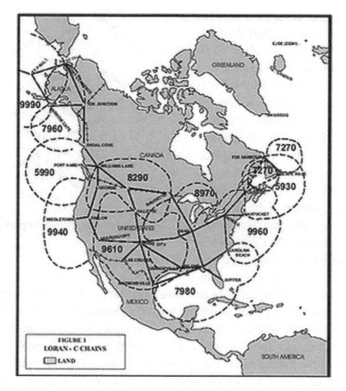

Figure 15-1:
Locations of LORAN transmitters. Only recently were these transmitters taken out of service.

Coordinated Universal Time (UTC), providing the most accurate timing information we have, even today.

GPS: Electronic Solution #2

While LORAN was a great success, it was not without its limitations, which were briefly described above. The major functional limitations were the resolution (a few hundred meters was good, but not as good as desired) and the lack of complete coverage of all the open ocean. Early in the era of satellites (starting in the 1960s), various systems were proposed to solve these problems with satellites, but the issue of cost was always too big. This is another area where Moore's Law (see Chapter 4) eventually came to the rescue.

To solve the problem, a network of satellites would be needed, using anywhere from about 10 to over 70 satellites to cover the entire surface of the Earth, depending on the altitude of the satellites. There's always a huge trade-off issue here: satellites which orbit closer to the Earth are less expensive to design, build and launch, but you need many

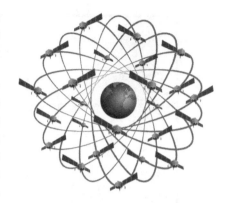

Figure 15-2:
GPS satellites in orbit around our planet.

more since their signal does not cover as much of the Earth's surface, and their orbital period is much shorter. Conversely, satellites which orbit farther from the Earth are more expensive to design, build and launch, but you don't need as many of them, since their signal covers more of the Earth's surface, and their orbital period is much longer.

Each of these satellites would have to have multiple atomic clocks for an extremely high-precision time reference, a high-frequency transmitter, and other sophisticated electronics. At first these items were prohibitively expensive, but soon the onward progress of Moore's Law brought the cost of the electronics down enough that the U.S. military approved and initiated the Global Positioning System (GPS) project in the early 1970s. It took many years for the very complex system to be designed, and the satellites were launched from 1989 through 1994.

The GPS network includes 24 operational satellites with 3 backups. Their orbital trajectories and altitudes are designed such that there should always be at least 4 of them in view from anywhere on the Earth. Each GPS satellite transmits a unique timing signal, derived from four on-board atomic clocks and accurate to within a few microseconds. Each GPS satellite sends a navigation message at the very low rate of 50 bits per second. Since a complete navigation message is 30 seconds long, it usually takes a GPS receiver about a full minute to pick up the necessary timing signals and navigation messages from the four GPS satellites necessary to fix the position of the GPS receiver.

GPS RECEIVERS

The cost reduction of the receivers has been a true miracle, in the author's experience. The GPSystem was available in the mid-90s, and many non-military consumers immediately desired its advantages. However, receivers were bulky (many pounds or kilos) and cost several thousand dollars, so they were mainly used in applications where they were essential. But again, the steady march of Moore's Law has reduced the size and cost of these receivers, where now the main function of a GPS receiver is simply integrated into many models of cell phones!

The exact orbit of each GPS satellite is well known and highly predictable, which makes each GPS satellite a known (though moving) reference point. And because each GPS satellite has its own unique timing signal, a GPS receiver out in the open has all the references it needs: at least four different location references and four different timing references. By combining these all together, a GPS receiver is able to determine its exact position to within a few meters. It also has extremely accurate time information, though this is generally only used to determine the position of the receiver.

GPS satellites orbit at an altitude of about 12,548 miles (20,200 km), which means their orbital period is just 2 minutes short of 12 hours. Because they are so far away from the surface of the Earth, their signal is heavily attenuated, and extremely sensitive receivers are required. One vendor of GPS receivers advertises the sensitivity of their receiver to be -158 dBm, which means it can detect GPS satellite signals as weak as 0.1 attoWatt (0.1 x 10^{-18} Watts). It is difficult to put this incredibly tiny amount of power into perspective, but it's worth a try.

One way of looking at this figure of 0.1 attoWatt might be to compare it to physical things. An ant crawling over a stick expends an extremely small amount of energy doing so. How does that energy compare to 0.1 attoWatt? Let's assume an ant weighs 1 g (0.035 oz), and that the stick is 2 mm (0.0787 inches) high, and that the ant climbs this height in 1 second. Assuming no losses, this means the ant expends 14.45 μhorsepower in climbing this stick. And since 1 horse-power = 745.7 Watts, this means the ant has expended:

14.45 μhorsepower * 745.7 Watts/horsepower = 0.0108 Watts

in climbing over this small stick. Yet that tiny amount of power is 10^{17} (100,000,000,000,000,000) times *greater* than the amount of power that a GPS receiver can successfully receive.

Let's next try using an example of electrical power with which we are familiar. One of the smallest amounts of electrical power we are familiar with is watch batteries. These batteries are typically rated at about 1.5 Volts and 20 mAh (20 mAh means it can supply 20 mil-liAmps for up to 1 hour, or 10 mA for 2 hours, etc.) So if one of these button-cell batteries lasts for about 3 years, this means it is supplying current for:

3 years * 365 days/year * 24 hours/day = 26,280 hours

which means it supplies an average current of:

20 mAh / 26,280 hours = 0.76 μAmps

which, in turn, means it supplies an average power of:

1.5 Volts * 0.76 μAmps = 1.14 μWatts.

This is a very tiny amount of power, yet it is 10^{13} (10,000,000,000,000) times *greater* than the amount of power that a GPS receiver can suc-cessfully receive. This is truly an amazing miracle of modern electronics.

GPS UNITS

The term "GPS" has become synonymous with the new and extremely popular GPS receivers which come with built-in maps. In fact, if you search the Internet for the term "GPS", the first hits you'll get are for these units from manufacturers such as DeLorme, Garmin, Magellan, NAVIGON, and many others. These units contain a GPS receiver, a microprocessor (as does almost anything electronic these days), and memory containing mapping information. The micropro-cessor runs specialized software which takes the mapping information, the GPS receiver information (telling the unit where the receiver is at each moment), and shows this information on the display of the unit. It is a great example of a highly complicated and sophisticated electronic device which has been made into a wonderfully useful navigation tool.

Another feature which is becoming increasingly popular with these units is traffic information. Knowing where you are is very helpful, but knowing the fastest way to get to your destination—depending on the traffic—is even better! These GPS units include another radio receiver which is tuned to your traffic information provider (often a separate service), and from that service it receives traffic information which the GPS unit then integrates into the mapping and route information. Then the friendly, imperturbable voice is able to tell you which turns to make, based on the actual conditions of the roads where you want to go. Again, this is a classic example of a case where extremely complicated functions have all been integrated into a unit which has been made about as small and light as practical, limited primarily by a practical size of the display.

LIMITATIONS

The major limitations of today's GPS units are the time it takes them to "wake up" and find where they are, the maps they contain, the battery life, and where they will and won't work.

As explained before, the wake-up time is mostly a function of the very slow 50 bits/second navigational information coming from the GPS satellites, so this time is not likely to improve in the near future. It would take an upgrade to each of these satellites, and that is simply not practical.

The maps contained in a GPS unit are mostly a function of what you pay and where you buy it. If you buy it in Europe, it will probably contain European maps. Getting detailed maps of other countries is usually something you do after you buy the unit. And how many maps it can hold is mainly a function of the memory it has, so this will certainly improve in the years to come. Many maps are available for download, at prices ranging from free to over $100 (US), depending on factors such as the level of detail, the accuracy, and the special features included in the maps.

Battery life is a bugaboo with all portable electronics. We'd all love for our cell phones, our netbooks, our notebooks and our GPS units to go for years without a recharge, like our watches do, but this is not going to happen in the near future. The more functions we have on our portable electronic devices, the more power they take, and batteries just cannot keep up with all the features we love. GPS units now can only go for a few hours without being plugged in; this is not likely to change much in the years to come, since battery technology has *not* improved at the Moore's Law rate, and is not likely to do so.

Where GPS units are fully useful is primarily a function of the sensitivity of the GPS receiver and of the type of building or geological feature you're in. The farther inside a building you go, or the deeper you descend into the basement (or a cave), the less likely you'll get any GPS signals from the distant GPS satellites. The future will probably include detailed maps of the inside of some buildings, and GPS repeaters inside the buildings, so this limitation will probably improve with time. With GPS repeaters in subways and elevators, (much as with cell phones), we may someday find that our GPS units will work wherever we go!

Chapter Take-Aways

After reading this chapter, one thing I would truly hope each readers feels is a deep admiration for early seafaring explorers, as exhibited by some of the names of their ships: *Dauntless, Intrepid, Fearless*. Their accomplishments were nothing less than amazing.

Equally impressive is the sophistication and complexity of the hardware and software that makes a modern GPS receiver so extremely useful in travel. The (mostly) smooth integration of maps, streets, location, distance, velocity, and tourist information has made these units truly the traveler's friend. The fact that they work at all is fundamentally due to the massive investment in the network of GPS satellites orbiting our planet.

CHAPTER 16
GAINING THE PERFORMANCE EDGE—
AUTOMOBILES

The first automobiles were available with three kinds of power: internal combustion, steam, and electric. But none of them had any electronics, because the first electronic devices had not yet been invented. Not long after electronics became available in the form of radio, automobile owners soon found ways to welcome these advances into their cars. And it was not long after this that automotive manufacturers found ways to offer radios in their new models.

But remember that the first radios were AM only, and that the models for the home were rather large (see Figure 16-1). It was more than difficult to find a way to accommodate such units in an automobile. However, the demand was very evident, as several individuals successfully mounted their "portable" radios in their autos. By about 1927, radio manufacturers made models that were specifically advertised as "car radios". This was the first introduction of electronics into automobiles.

Today, electronics in automobiles do much more than give us an AM radio station to tune to. They provide several entertainment options, safety features, communication features, comfort features, and performance features. And all this in one of the most demanding applications, for automotive electronics must endure very high and very low temperatures, a wide range of humidity, and relatively

Figure 16-1:
An example of an early home radio. These units were commonly 20" (50 cm) wide by 10" (25 cm) deep by 25" (62 cm) high, and weighed about 20 lbs (9.1 kilos).

235

high levels of vibration, all of which can significantly reduce the lifetime of electronics.

Automotive Entertainment

Current automotive entertainment options, as always, have closely followed home entertainment options, with the exception that the viewing screen is significantly smaller. Eight-track tape players gave way to audio cassette players, which have now given way to CD players. And for those in the rear seat, DVD players, video game players, and Internet access is available. Becoming thornier are the issues of front-seat access—Internet access is a great option for the driver *if* the car is not moving. But that's an issue that will need to be worked out.

Most consumers would agree that one of the best improvements in automotive entertainment systems is the quality of the units that come with the car. Not too long ago, if you wanted a good entertainment system, you took out the one that came with the car and replaced it with a top-rated aftermarket system, and there were many businesses whose only service was to do just that. But these services have recently been declining, because automotive manufacturers are providing some excellent entertainment systems right from the factory.

Safety and Convenience Features

Several relatively recent safety features have greatly improved the safety of automobiles, including anti-lock braking systems (ABS), air-bag systems (more recently known as supplemental restraint systems or SRS), outside temperature warnings, adaptive cruise control, backup camera and sensors, infrared night vision, adaptive highbeams, traction control systems, lane departure warning systems, tire pressure monitoring systems, electronic stability control, adaptive wipers, driver wariness monitors, and even automated parking systems.

All of these systems involve a significant amount of electronics, and essentially all of them use a microcontroller to handle the input from sensors and provide the necessary output to the actuators. All of them have electronic sensors to sense specific conditions. These sensors for monitoring deceleration (detecting a crash), temperature, vehicle speed and tire rotational speed (for detecting a difference indicating a loss of traction), light, infrared vision, fixed obstacles (for backup warnings and automated parking), driver head angle (for monitoring drowsiness), moisture (for adaptive wiper systems), and tire pressure—used to be prohibitively expensive. Highly innovative integrated circuit design has made it possible to combine the sensor, the signal processing, and the control electronics all on the same chip, reducing the costs by 10s to 100s of times. It is this constant reduction of costs, coupled by constant increases in features, which is possible with electronics due to the constant march of Moore's Law (see Chapter 4).

Chapter 9 discusses how computers work, and one section talks about inputs and outputs. For computers, we are comfortable that a keyboard and a mouse are input devices, and that a printer and a monitor are output devices. But microprocessors, around which computers are built, are actually much more versatile than this, and microcontrollers are simply embedded microprocessors (embedded meaning you can't easily program them yourself, and there is no keyboard, mouse, and sometimes no display). Microcontrollers use inputs from sensors, and send output to actuators. The simplest example of this is a modern thermostat: its inputs are the temperature in the home and the desired temperature (set point); its outputs turn on either the cooling or the heating, depending on which is needed.

Today's automobile has an average of about 25 microcontrollers, depending on the features of the vehicle. Each of these microcontrollers has more computing power than early desktop computers. And each makes possible a system with optimized performance and minimal cost.

Communication Features

Early automobiles were essentially a traveling cocoon—someone outside the automobile could not communicate to anyone inside, nor could someone inside communicate with anyone outside (other than to roll down the window and yell!). The cell phone (see Chapter 11) changed all that, but designers recognized the possibility of doing still more. Hands-free talking has both simplified using a cell phone while driving as well as made it safer.

Bluetooth devices have simplified synchronizing portable devices with the automobile, including MP3 players, cell phones, and notebook computers. A bit of flavor is relevant here.

Why the name "Bluetooth"? The story is that the company who invented the Bluetooth technology (Ericsson, of Denmark) intended to replace cables interconnecting devices over short ranges. Looking through their country's history, they recalled King Harald I (c. 935 AD), who was successful in uniting many of the Danish tribes into a single kingdom. Apparently he was very fond of blueberries, to the point that his teeth took on a permanent bluish tint, so he earned the nickname Bluetooth. And since this new technology would unite many separate devices, it seemed appropriate, and it has been quite the market success.

The Bluetooth protocol is intended to have a range limited to about 10 meters, and permits automatic creation of piconets, meaning very small networks of interconnected devices. When so enabled, Bluetooth devices detect each other, create the communication link, and begin to exchange data, all without any user interaction. This is very convenient for connecting portable devices to the entertainment system of an automobile.

Comfort Features

Modern comfort features include climate control, seat position personalization, and ride comfort adjustment. Again, each of these features depends on sensing the status of systems in the automobile and responding appropriately to these conditions. It is as though the vehicle recognizes who we are and adjusts things accordingly, and that is just what designers are working toward. I would predict that the future of automotive comfort would lie in this area of personalization.

Performance Features

All of the above features are wonderful and amazing, and all have been enabled by inexpensive and high-performance electronics. But in this author's opinion, they pale in comparison to the performance features.

Performance features nearly always make their debut in the racing technology, where teams of technicians do everything legal and feasible to extract the maximum performance from a given set of limitations. Advances in performance have come from improvements to the fuel system, ignition system, lubrication system, suspension system, drive system, timing system, and a few miscellaneous improvements.

FUEL SYSTEM IMPROVEMENTS

The first automotive fuel systems consisted primarily of a pump to take the fuel from the tank and vaporize it prior to entering the cylinders of the engine. While these systems were basically functional, they were also very finicky, and often did not work well. In reality, the performance of a given fuel in an internal-combustion engine is a function of a variety of factors, and the interaction of these factors is also quite complex. Some of these factors include the temperature of the air, the humidity, the elevation (and thus the density of the air), other contaminants in the air, the temperature of the fuel, the oxygen content of the fuel, the purity of the fuel, and the status of the fuel system overall.

Before electronic fuel injection (EFI) systems were developed for racing, internal combustion engines were fed fuel through carburetors, whose job was to mix the fuel with air as optimally as possible. But carburetors cannot respond to the many different variables, except when they're tuned, and tuning them for each specific condition is not something consumers would ever take the time to do. So fuel systems using carburetors stumbled along as well as they could (and some very impressive results were obtained!) But when EFI came along, racing quickly adopted it, and significant gains were posted. Eventually these performance gains trickled down to consumer automobiles, and today, it is doubtful that any new cars use carburetors.

EFI also allows precise control over the amount and timing of fuel being injected into the cylinders. This single improvement over the carburetor was sufficient to allow EFI to eventually replace the carburetor. But one of the most useful advantages of EFI is its ability to sense and respond to the actual conditions under which the engine is operating. Today's engines sense all of the parameters mentioned previously (or very close to it), and this information is passed along to the EFI system. The microcontroller then executes the code that corresponds to those unique conditions, enabling significant performance gains over a carbureted system.

In addition to performance gains, EFI systems also enable very significant improvements in emission reduction and mileage improvement. The combustion of fuel in internal combustion engines has been studied in great detail for decades. Engineers and scientists know well what the optimal conditions are for burning the fuel, leaving as few byproducts as possible and extracting the maximum amount of energy possible. Since the fuel combustion conditions are constantly changing as a function of the many parameters mentioned previously, only an electronic system is able to respond to each parameter and adjust the controls accordingly. Today's EFI systems do this remarkably well, and there will yet be more improvements in these systems.

IGNITION SYSTEM IMPROVEMENTS

The ignition system of an automobile is closely tied to its fuel system. It is the (mostly) electrical and electronic system that allows an engine to be turned on, turned off, and ignite the fuel in each cylinder at the optimal time.

The turning on and off of an engine is accomplished electrically (for non-diesel engines), as it has been since the invention of the engine. Each cylinder has a spark plug, whose purpose is to take the 20,000–30,000 Volts fed into the spark plug and convert it into a small spark inside the cylinder. Assuming the cylinder contains the proper mixture of fuel and air, the spark causes the fuel/air mixture to ignite. If the spark plug is deprived of this high-voltage jolt, the fuel does not ignite and the engine dies. So, turning on and off an engine is as simple as turning on and off this high-voltage ignition jolt that goes to each cylinder.

The exact moment when this high-voltage spark occurs is key in the performance of an engine. Assuming the EFI system is working properly and there is a perfect mixture of fuel and air in each cylinder, the only remaining variable is this spark—its exact timing and magnitude. Providing this carefully controlled spark is the job of the electronic ignition system.

Prior to electronic ignition systems, the timing of this spark was controlled by a rotor which was mechanically attached to the crankshaft. This worked well under a single condition, but as variables changed, the timing needed to be adjusted, and it had to be done mechanically. Today's electronic ignition systems take the same information that the fuel system has obtained about all the relevant parameters, and use this information to adjust the timing dynamically, providing another way of producing optimal performance from the engine, under a wide range of conditions.

As an aside, it can seem impressive that electronic ignition systems can provide these carefully controlled sparks at a rate of 12,000 times per minute (assuming a 6-cylinder, 4-stroke engine running at 4,000 rpm), but this is only 200 times per second, and electronic systems can do this MUCH faster—even billions of times per second. But what this author finds impressive is that an automobile engine running at 4,000 rpm has pistons which are going up and down 66.7 times every second, or once up and down every 15 milliseconds—that's pretty amazing! And what about racing engines capable of 12,000 rpm—this means those pistons are going up and down 200 times per second, or once every 5 ms! That is incredibly fast for something so mechanical and so tied to the laws of acceleration and mass. (Electronic circuits are not really tied to acceleration and mass).

LUBRICATION SYSTEM IMPROVEMENTS

The basic lubrication system of an internal combustion system remained essentially unchanged for decades. Recent years have seen several improvements, but as far as this author is aware, only one of these improvements can be attributed to electronics.

Knowing when to change your oil is very important in maintaining an engine. If you go too long between changes, you invite excessive wear and corrosion. If you change it too soon, you simply waste the extra time the oil could have served adequately. While optimal timing for oil changes is best, the biggest problem faced by most automobile owners is doing it at all.

Most of us would prefer to simply use our automobiles and not worry one second about how all their systems actually work. We'd like oil changes to be something like fuel—the car tells us when it's time to refill the tank. So many modern automobiles do just that—they keep track of how many miles have been driven (or how many months have elapsed) since the last oil change, as well as the temperature of the oil (another major factor in how long oil lasts), and they turn on a warning light when it's time to change it again. After the oil is changed, the service provider resets the light and the owner is good to go until the next change is warranted.

It is arguable whether this oil change warning light turns on at the *optimal* time, but it is for sure that not changing the oil at all is terrible, and that following the recommendation of this oil-change indicator will prevent owners from forgetting to do it at all.

SUSPENSION SYSTEM IMPROVEMENTS

A suspension system, to most of us, is simply that part of our automobile which determines how smooth or how sporty the ride feels. But as far as safety and performance goes, a suspension system is another key contributor. Keeping the tires on the road as much as possible adds greatly to enabling greater control and performance in all driving conditions.

The suspension system of an automobile consists of many pieces, but only two will be discussed here: the shock absorbers and the springs. The springs are to cushion the bumps, and they do the job quite well. It is fairly easy to envision their operation by recalling the difference you've probably noticed between running with dress shoes on (hard soles), versus running with good running shoes (on cushion soles)—the cushion soles do much to reduce the shock of each bounce.

While springs do much to absorb the bumps encountered by the automobile, they suffer from a problem: they bounce back. In this sense, the springs of the automotive suspension system are much like a ball: if you drop it, it will bounce back. So riding in an automobile with only springs for its suspension system would be somewhat like sitting on top of a large, inflated ball—every time you encountered a bump, the car would begin bouncing up and down, and it would take a few seconds for this bouncing to stop. That would become quite annoying in a very short time!

The shock absorbers are designed to stop this bouncing. Their job is to move quickly when a bump is encountered, but to move slowly when the spring tries to bounce back. The load rating of the shock absorbers must be closely matched to the spring, so that an optimal ride is obtained.

But there is a range of rides considered optimal, depending on the type of vehicle. Large luxury sedans are supposed to have a very smooth ride, while high-performance sports cars are supposed to have a much stiffer ride. And there is a range of rides in between these extremes.

In addition to the ride delivered by the suspension system is the desire to keep the tires fully in contact with the pavement at all times. Clearly, each time a tire bounces up it loses its adhesion to the road surface, which reduces control. This effort to keep the tires in contact with the road surface is another variable in the suspension control system.

So, today's top-of-the-line suspension systems are electronically adjustable! You just select the type of performance you desire, and the response of the suspension system is adjusted to provide the kind of ride and road-surface adhesion that is desired. Since the performance of the vehicle is adjusted in only a few seconds (or less), this kind of control has become quite popular in many models of automobiles.

One of the ways that this electronically adjustable ride is made possible is through the use of a special type of fluid in special shock absorbers. *Magnetorheological* fluids are fluids that have special materials in the fluid. These materials respond to magnetic fields in such a way that the flow properties (*rheological* properties) are modified. Such materials are relatively recent, and their application as shock absorbers is also quite recent. So, with some relatively simple electronic controls, you can make the fluid in the shock absorbers flow *more* readily, which softens the ride; or you can make the fluid in the shock absorbers flow *less* readily, which stiffens the ride. And these changes can be accomplished in only a few seconds.

DRIVE SYSTEM IMPROVEMENTS

The ideal drive system would sense the road conditions and how well each wheel is gripping the road, then provide drive only to those wheels which are providing the necessary grip. The amazing thing is that such drive systems are available today! They're not perfect yet, but the improvement they provide is very significant.

This is particularly true with 4WD or AWD drive trains, where traction control systems are able to keep such vehicles from becoming stuck as soon as the first wheel loses traction. While the drive systems are primarily mechanical, the sensors and controls that engage these traction control mechanisms are electronic.

Traction control systems make getting stuck less likely, but they also provide one other very desirable feature: control on corners and slippery roads. Humans are generally not able to sense what is happening to the automobile on corners or slippery roads until it is too late and the vehicle is out of control. Fortunately, electronics can come to the aid here. By monitoring each wheel and only applying drive to those which are gripping the road properly, it is much easier for drivers to stay in control in difficult road conditions.

MISCELLANEOUS IMPROVEMENTS

There are many other automotive improvements which could be discussed, but not all of them are interesting, at least to this author. But adaptive cruise control and automated parking are two that seem futuristic enough that I believe they deserve mention. And they're not that futuristic—they've been demonstrated on concept cars, and are available on some top-end automobiles.

Adaptive cruise control is one step closer to taking the human driver out of the loop. This can be scary to most people, but we should remember that humans are the #1 cause of automobile crashes, and it's usually because they are distracted. Microcontrollers don't get distracted, and they never get tired of watching the road. Ordinary cruise control keeps your automobile traveling at the same speed, regardless of uphill or downhill roads; this has greatly relieved drivers on long trips from the continuous task of looking at the speedometer and changing the position of the accelerator to stay at the same velocity. Adaptive cruise control takes it one step further, by watching cars in front and slowing down automatically when the distance becomes too small. And all this takes is a radar system which can sense the position and velocity of cars in front and to the side of your car. While this sounds easy, in reality it is quite far from it. Radar has always been quite expensive, and while it is very effective, it has only been justified where safety is a major concern, as in airplanes. It is truly a great advance that radar systems for this application have become affordable and small enough to be practical for automobiles.

Automated parking is a lot of fun to watch! Just choose your favorite Internet video website and search for "automated parking"—you'll want to go out and buy such an automobile today! This technology uses some of the same sensors used for adaptive cruise control for monitoring the position of nearby vehicles, and provides the steering controls automatically. All the driver has to do is to keep the vehicle moving slowly—the steering and collision avoidance are all taken care of by the automated parking system. And some systems are completely automated—just position the automobile near where you wish to parallel park, push the button, and the entire process is done completely for you!

The Future of Automotive Technology

The best way to learn about the future of automotive technology is to read about what race cars are doing to improve their performance, or to read about concept cars put out by various automotive manufacturers, or to read about the DARPA Grand Challenge for driverless vehicles. These technologies are unaffordable for today's general consumer, but for those technologies which are based on electronics, time and the onward march of Moore's Law will eventually make them affordable.

What are some of these technologies? Certainly the all-electric car will eventually become a reality, and is bound to be a success ultimately. All-weather lighting systems are being designed, which adapt road lighting to fog, rain, snow, or other weather conditions. Heads-up displays (transparent displays projected in front of the windshield for the driver only) are again in the news, as well as adaptive night vision (allowing you to see that deer in time to avoid it). Rapid and automatic integration of a smart phone will allow your cell phone to become the car phone, the Internet access link, the video player, and the automobile information source. This will allow the driver to adjust the sound system and enter the route plan from the comfort of their home. Since this is done over the cell phone system, it can be done at a considerable distance from the car.

Organic Light-Emitting Diode (O-LED) displays have made their debut in concept cars and will work their way down to the high-volume cars. O-LED displays provide outstanding contrast while requiring less power than even liquid-crystal displays (LCDs).

Will the time ever come that automobiles drive themselves? In my opinion, yes. There are several motivating factors for this development, and the technologies being developed for the DARPA Grand Challenge for driverless vehicles are truly amazing. The main motivating factor, as mentioned previously, is simply that humans are the main cause of the massive number of traffic fatalities and injuries. Whether distracted or substance-impaired, drivers are an extremely weak link in the world-wide traffic system. But the reason we let all of these impaired humans (including myself) keep driving is that we have no better alternative. But when the day arrives that computers can be shown to outperform humans (and that day *will* arrive), it seems inevitable that the transition will begin. It will take vehicles with amazing navigation systems, sensors (including radar), maps, communications (including with other cars), and software, as well as intelligent roads (providing information on accidents, weather, traffic flow, etc.), but it seems inevitable that it will happen. It will not happen in 10 years, but I believe that within 20 years, we will see some vehicles operating autonomously in limited circuits. And from there, it's only a matter of time.

Chapter Take-Aways

In Chapter 4, a bullet list was given of what today's automobile would be like if it had been improved at the same rate that Moore's Law has allowed integrated circuits to be improved. Some items on that list are ludicrous, and will surely never happen, but there is no doubt that automobiles will continue to be improved. Many of the improvements to automobiles over the past few decades have been mainly due to the integration of electronics into the various systems of the vehicles. These will certainly continue.

And gone are the days when many systems of the automobile are controlled directly from inputs from the driver (steering wheel, accelerator, brake). Instead, much of the car is driven by software: if you try to turn the steering wheel very sharply while traveling at a high velocity, the software simply prevents that from happening (that prevents *overcorrection*, a major cause of single-vehicle accidents). If you try to accelerate or decelerate too quickly, the system simply takes over and prevents a peel-out (via traction control systems) or a loss-of-control quick stop (via anti-lock braking systems). So, in some ways, we have already given some control of our vehicles over to the electronics—and our vehicles are safer because of it.

So, in some modern high-end vehicles, when you use the accelerator to tell the car to quickly accelerate, or the steering wheel to quickly turn left, or the brakes to quickly stop, you do not actually control the car. Instead, you send information to the computers which control the car, telling these computers that you would *like* to accelerate, turn or stop, and the *computers* decide if it's safe to do what you'd like. If it isn't, your car won't do it! So who do you *really* think is in control of that automobile?!

Chapter 17
Devices on the Horizon and Beyond

Over the past 100 years, we have seen electronics grow from a tiny, newborn infant technology to one which has revolutionized the commercial world, giving us a plethora of consumer, military, and industrial products which would never have been possible otherwise. As outlined in Chapters 2, 3, and 4, this has been possible because of the vacuum tube, the transistor, and the integrated circuit. For the last 50 years, there has been nothing as revolutionary and new as these three advances, yet improvements have continued. Can this continue indefinitely?

Certainly not. Moore's Law (which is really just an observation—there is no real "law" here) only holds true as long as the integrated circuits industry finds ways to overcome increasingly difficult obstacles. The demise of our ability to continually shrink transistors in integrated circuits has been forecasted many times, yet each time some way has been found to continue the onward march. But we are not far away from a dead end; all the experts agree on that. The only thing they disagree on is how many more years we have left, and which way to go next. Twenty years ago, the experts tended to agree that there were only about ten years left; ten years ago, the prediction was about the same. So it should come as no surprise that today's predictions are about the same: ten more years. But this time the barriers are bigger than ever—it will take some amazing advances just to get those ten more years in the forecast.

But even if we do not have ten more years of Moore's Law in integrated circuits, this does not mean that all progress in computers, cell phones, entertainment, and so forth, will cease. For many years, researchers have been looking into many different devices to potentially take the place of the MOSFET transistor, which is the workhorse of today's integrated circuits. These potential future devices have ranged from the rudimentary to the wildly exotic, and the stakes are massive. The electronics industry is one of the largest industries in the world, so successfully coming up with its successor is of great interest to many people and companies. In this chapter, we will take a look at a few of these.

Carbon Nanotube Transistors

Amazing things have happened with carbon in the past few decades. Prior to these discoveries, carbon had basically two main allotropes (different physical arrangements of the atoms of a given material): crystalline, which we know as diamond, and hexagonal sheets, which we know as graphite. One of the discoveries, for which the discoverers eventually received the Nobel prize in 1996, was an allotrope called "fullerenes" (named after Buckminster Fuller, who designed geodesic domes)—so named because their shape was very similar to Fuller's domes. Their similarity to the shape of a soccer ball also lent to their being called "buckyballs". This allotrope of carbon was found to have some very unique properties, which are still being investigated. Continued research into this unique arrangement of carbon atoms eventually produced carbon nanotubes, which also have some very unique properties.

These tiny tubes of carbon are only a few nanometers in diameter, but can be grown to be several centimeters in length, or much shorter. In an arrangement of nanotubes which are only a few times longer than their diameter, carbon nanotubes have shown switching characteristics. This means they could be used to make a device which could switch like a transistor. If researchers are successful in making switching devices based on carbon nanotubes, their advantages would include being up to 100 times smaller than today's MOSFET transistors, using up to 100 times less power, and being more reliable. Today such devices are only experimental, but results so far have been promising. As usual, the main challenges are making them practical in large quantities, meaning cost, manufacturability and reliability.

Photonics

So, aside from the catchy name, what is the field of photonics? One of the best ways to understand this field is to compare and contrast it with a field we have grown accustomed to, and on which this book is focused: electronics. The field of electronics concerns itself with devices whose operation is best understood by describing their operation in terms of *electrons*, those quantum particles of negative charge which surround the nuclei of most atoms. Electronic devices include all those discussed in this book, particularly in Chapters 2-4: the vacuum tube, the transistor, and the integrated circuit.

Photonics concerns itself with devices whose operation is best understood by describing their operation in terms of photons, those quantum units of light that first established the field of "particles" which possess properties of *both* particles and waves. There are many devices today which possess properties of both electronics and photonics, which include laser diodes, solid-state lasers, light-emitting diodes (LEDs), avalanche photodiodes, photoresistors, and others.

Much research has been poured into devices which are purely photonic, allowing photons to be created, turned on and off with light, switched from one output to another, and basically to be controlled in many of the ways we presently control electrons through electronic devices. Some of these devices are nearing becoming practical realities. When all the necessary functions can be integrated together on a single substrate, we will have the rudiments of a fully photonic computer. Photonic computers will operate much faster than today's computers, and use much less power.

Quantum Computing

The extremely strange world of quantum mechanics allows us to do things that are quite out of the ordinary in the normal world we live in, including having switching devices which can be in both states at once, as well as quantum entanglement which allows digital communication over any distance in zero time. Quantum bits are termed *qubits*, and while research in this area is still very new, laboratory results have been promising.

It is somewhat sobering to grasp the capability of even a relatively small qubit-based computer. Researchers claim that a qubit-based computer could solve in seconds some problems that would take today's fastest supercomputers many years.

Practical devices based on qubits are many years away, but due to their promise, it is likely that researchers will continue to study them and that improvements and progress will come. Presently they only operate at cryogenic temperatures (liquid nitrogen or even colder). They are not likely to be practical in the next decade.

Atomic-Scale Computing

This area, like the one before, is still very new, and practical devices are still very far away. But reports indicate that researchers have been able to fabricate in their labs logic gates with as few as seven atoms. For comparison, today's transistors are about 100 x 100 nm, at a depth of a few nanometers, giving a volume of:

100 nm x 100 nm x 10 nm = 100,000 nm^3

which, at a spacing of about 2.5 silicon atoms per nm, gives a volume of:

250 atoms x 250 atoms x 25 atoms = 1,562,500 atoms.

A logic gate is made up of about 15 transistors, so today's technology needs about:

1,562,500 atoms/transistor x 15 transistors = *23,437,500 atoms*

for a single logic gate. So, our final comparison is that an atomic-level logic gate would be about:

23,437,500 atoms / 7 atoms ≈ 3,348,214 times smaller!

We can also generally assume that such devices would use much less power than their much larger counterparts, although not exactly at the same scale of 3,348,214 times less. A reasonable guess would be several thousands of times less power. And what about switching speed? In today's transistors, switching speed is very closely related to the size of the transistor, and it is reasonable to expect that this would be true to some degree for these atomic-level devices. If they were even just 50 times faster, this would be a HUGE improvement; but given how much smaller they are, it is not unreasonable to hope that they could be more than 100 times faster—or more.

Biological Computing

This area has certainly been interesting for this author to follow. Over the past 20 years or so, there have been numerous reports on break-throughs in making small logic gates or switching devices out of some biological element, including neurons, tissue, and laboratory-grown cultures. These have the potential of providing some advantages such as lower power consumption and lower cost "manufacturing", but cells are actually much larger than today's transistors, so at the size of a cell, they won't provide a size advantage. A typical cell in the human body (such as a red blood cell) is about 10 μm in diameter; today's transistors are about 100 x 100 nm.

But there is one size advantage they could have, and that is their third dimension. Today's transistors are commonly all in a single layer on the integrated circuit; they are not made in multiple layers (yet). Biological cells lend themselves very well to a third dimension, and if a way can be found to take advantage of this fact, there could indeed be some size advantage with them.

The idea of biological computing gives rise to some very strange possibilities, including a physical computer virus, diseases, cell replacement (they would still have a finite lifetime), and handling of waste products (they must metabolize to stay alive, which means they produce waste products). All this could lead to some very interesting fields if biological computing becomes a reality.

In the Meantime

Each of the technologies mentioned above is very new, with the possible exception of photonics. All are being researched, but none is likely to be available in the next 10 years to take the place of the MOSFET transistors in use today. So what are semiconductor manufacturers doing about this? A great deal! The life of their companies is on the line, and they have been working on this for decades. Most semiconductor manufacturers around the world joined together many years ago to regularly produce a document known as the International Technology Roadmap for Semiconductors. This roadmap is an incredibly detailed document; the Executive Summary of the 2009 version is 95 pages long! It describes the huge problems the industry is facing and what is being studied or tried for solving these problems. This document has been renewed regularly, and each edition makes it sound like the problems are so huge they'll never be overcome. It categorizes problems four ways: 1) those with manufacturable solutions already existing and being optimized; 2) those with known manufacturable solutions; 3) those with known interim solutions; and 4) those with no known manufacturable solutions. It shows plans to reach sizes of 18 nm in 2015, 10 nm in 2020, and 6.3 nm in 2024, but a detailed examination shows that many pieces of these plans are category 4 challenges—no known manufacturable solutions exist.

Tens of thousands of people are working on these challenges, with companies spending billions of dollars each year. The financial incentive is there—the industry is huge. There is no way that Moore's Law will simply die—it will just get harder and harder to keep something like it going. Surely the rate of improvement will decline, as it has in the past decade. And eventually it will grind to a halt, at least in some aspects. But judging by the past, I would not bet that it will grind to a halt in the next 15 years; there is a very long record of success in keeping the progress going.

Chapter Take-Aways

The future of electronics is bound to include some very notable new products and applications. Hold on to your pocketbook and to your favorite electronic gadget! In the meantime, the entire civilized world has become greatly enamored of the success of these products. The advances seen in the past 50 years will probably pale in comparison to what will happen in the future. Features and functions will continue to increase; costs will continue to decline. Breakthroughs will continue to be announced, and Nobel prizes will be awarded to those who make fundamental advances.

When the transistor was first made practical in the early 1950s, it was expensive and rare. Today hundreds of trillions of transistors are made every *day*. And the ways these transistors are applied will continue to grow, amazing all who follow the industry. So enjoy all those electronic gadgets you have, and continue to watch for more!

INDEX

Index

Index

ABOUT THE AUTHOR

Barry is currently a full professor of Information Technology at Brigham Young University, (BYU) in Provo, UT. He has been in the teaching profession for 26 years, all at the post-secondary level. He has varied professional experience, including 7 years as a design engineer with IBM, 6 summers as a faculty intern, and several involvements in consulting. He has recently been a founding partner of Millenniata, Inc., a company which has developed an optical disc for permanent data storage.

His experience with IBM was in the design of integrated circuits and circuit boards for ½-inch tape, where he became very familiar with the details of storing digital data. His summer internships have been with Bell Labs, Larson-Davis, and Micron Technology, in areas of integrated circuit testing and electronic measurement of physical properties of nonconductive materials. His consulting has been in areas of power management, electronic manufacturing, and combustion forensics. He has four patents, 20 patents pending, and has published over 50 papers and two books.

His academic degrees include a BS in Electronics Engineering Technology and an MS in Manufacturing and Electronics from BYU-Provo (Utah), and a PhD in Occupational and Adult Education from Utah State University (Logan, Utah). His publications are in the areas of electronic applications, long-term data storage, and technical education.

Barry continues to teach college classes in digital communication, data storage, electronic manufacturing, and computers, and has won numerous teaching awards and one technology transfer award. He is married to the former Catherine Page; they have four children and seven grandchildren.

A CATALOG OF SELECTED
DOVER BOOKS
IN ALL FIELDS OF INTEREST

A CATALOG OF SELECTED DOVER BOOKS IN ALL FIELDS OF INTEREST

100 BEST-LOVED POEMS, Edited by Philip Smith. "The Passionate Shepherd to His Love," "Shall I compare thee to a summer's day?" "Death, be not proud," "The Raven," "The Road Not Taken," plus works by Blake, Wordsworth, Byron, Shelley, Keats, many others. 96pp. $5\frac{3}{16} \times 8\frac{1}{4}$. 0-486-28553-7

100 SMALL HOUSES OF THE THIRTIES, Brown-Blodgett Company. Exterior photographs and floor plans for 100 charming structures. Illustrations of models accompanied by descriptions of interiors, color schemes, closet space, and other amenities. 200 illustrations. 112pp. $8\frac{3}{8} \times 11$. 0-486-44131-8

1000 TURN-OF-THE-CENTURY HOUSES: With Illustrations and Floor Plans, Herbert C. Chivers. Reproduced from a rare edition, this showcase of homes ranges from cottages and bungalows to sprawling mansions. Each house is meticulously illustrated and accompanied by complete floor plans. 256pp. $9\frac{3}{8} \times 12\frac{1}{4}$. 0-486-45596-3

101 GREAT AMERICAN POEMS, Edited by The American Poetry & Literacy Project. Rich treasury of verse from the 19th and 20th centuries includes works by Edgar Allan Poe, Robert Frost, Walt Whitman, Langston Hughes, Emily Dickinson, T. S. Eliot, other notables. 96pp. $5\frac{3}{16} \times 8\frac{1}{4}$. 0-486-40158-8

101 GREAT SAMURAI PRINTS, Utagawa Kuniyoshi. Kuniyoshi was a master of the warrior woodblock print — and these 18th-century illustrations represent the pinnacle of his craft. Full-color portraits of renowned Japanese samurais pulse with movement, passion, and remarkably fine detail. 112pp. $8\frac{3}{8} \times 11$. 0-486-46523-3

ABC OF BALLET, Janet Grosser. Clearly worded, abundantly illustrated little guide defines basic ballet-related terms: arabesque, battement, pas de chat, relevé, sissonne, many others. Pronunciation guide included. Excellent primer. 48pp. $4\frac{3}{16} \times 5\frac{3}{4}$. 0-486-40871-X

ACCESSORIES OF DRESS: An Illustrated Encyclopedia, Katherine Lester and Bess Viola Oerke. Illustrations of hats, veils, wigs, cravats, shawls, shoes, gloves, and other accessories enhance an engaging commentary that reveals the humor and charm of the many-sided story of accessorized apparel. 644 figures and 59 plates. 608pp. $6\frac{1}{8} \times 9\frac{1}{4}$. 0-486-43378-1

ADVENTURES OF HUCKLEBERRY FINN, Mark Twain. Join Huck and Jim as their boyhood adventures along the Mississippi River lead them into a world of excitement, danger, and self-discovery. Humorous narrative, lyrical descriptions of the Mississippi valley, and memorable characters. 224pp. $5\frac{3}{16} \times 8\frac{1}{4}$. 0-486-28061-6

ALICE STARMORE'S BOOK OF FAIR ISLE KNITTING, Alice Starmore. A noted designer from the region of Scotland's Fair Isle explores the history and techniques of this distinctive, stranded-color knitting style and provides copious illustrated instructions for 14 original knitwear designs. 208pp. $8\frac{3}{8} \times 10\frac{7}{8}$. 0-486-47218-3

Browse over 9,000 books at www.doverpublications.com

ALICE'S ADVENTURES IN WONDERLAND, Lewis Carroll. Beloved classic about a little girl lost in a topsy-turvy land and her encounters with the White Rabbit, March Hare, Mad Hatter, Cheshire Cat, and other delightfully improbable characters. 42 illustrations by Sir John Tenniel. 96pp. 5³⁄₁₆ x 8¼. 0-486-27543-4

AMERICA'S LIGHTHOUSES: An Illustrated History, Francis Ross Holland. Profusely illustrated fact-filled survey of American lighthouses since 1716. Over 200 stations — East, Gulf, and West coasts, Great Lakes, Hawaii, Alaska, Puerto Rico, the Virgin Islands, and the Mississippi and St. Lawrence Rivers. 240pp. 8 x 10¾. 0-486-25576-X

AN ENCYCLOPEDIA OF THE VIOLIN, Alberto Bachmann. Translated by Frederick H. Martens. Introduction by Eugene Ysaye. First published in 1925, this renowned reference remains unsurpassed as a source of essential information, from construction and evolution to repertoire and technique. Includes a glossary and 73 illustrations. 496pp. 6⅛ x 9¼. 0-486-46618-3

ANIMALS: 1,419 Copyright-Free Illustrations of Mammals, Birds, Fish, Insects, etc., Selected by Jim Harter. Selected for its visual impact and ease of use, this outstanding collection of wood engravings presents over 1,000 species of animals in extremely lifelike poses. Includes mammals, birds, reptiles, amphibians, fish, insects, and other invertebrates. 284pp. 9 x 12. 0-486-23766-4

THE ANNALS, Tacitus. Translated by Alfred John Church and William Jackson Brodribb. This vital chronicle of Imperial Rome, written by the era's great historian, spans A.D. 14-68 and paints incisive psychological portraits of major figures, from Tiberius to Nero. 416pp. 5³⁄₁₆ x 8¼. 0-486-45236-0

ANTIGONE, Sophocles. Filled with passionate speeches and sensitive probing of moral and philosophical issues, this powerful and often-performed Greek drama reveals the grim fate that befalls the children of Oedipus. Footnotes. 64pp. 5³⁄₁₆ x 8 ¼. 0-486-27804-2

ART DECO DECORATIVE PATTERNS IN FULL COLOR, Christian Stoll. Reprinted from a rare 1910 portfolio, 160 sensuous and exotic images depict a breathtaking array of florals, geometrics, and abstracts — all elegant in their stark simplicity. 64pp. 8⅜ x 11. 0-486-44862-2

THE ARTHUR RACKHAM TREASURY: 86 Full-Color Illustrations, Arthur Rackham. Selected and Edited by Jeff A. Menges. A stunning treasury of 86 full-page plates span the famed English artist's career, from *Rip Van Winkle* (1905) to masterworks such as *Undine, A Midsummer Night's Dream,* and *Wind in the Willows* (1939). 96pp. 8⅜ x 11. 0-486-44685-9

THE AUTHENTIC GILBERT & SULLIVAN SONGBOOK, W. S. Gilbert and A. S. Sullivan. The most comprehensive collection available, this songbook includes selections from every one of Gilbert and Sullivan's light operas. Ninety-two numbers are presented uncut and unedited, and in their original keys. 410pp. 9 x 12. 0-486-23482-7

THE AWAKENING, Kate Chopin. First published in 1899, this controversial novel of a New Orleans wife's search for love outside a stifling marriage shocked readers. Today, it remains a first-rate narrative with superb characterization. New introductory Note. 128pp. 5³⁄₁₆ x 8¼. 0-486-27786-0

BASIC DRAWING, Louis Priscilla. Beginning with perspective, this commonsense manual progresses to the figure in movement, light and shade, anatomy, drapery, composition, trees and landscape, and outdoor sketching. Black-and-white illustrations throughout. 128pp. 8⅜ x 11. 0-486-45815-6

Browse over 9,000 books at www.doverpublications.com

THE BATTLES THAT CHANGED HISTORY, Fletcher Pratt. Historian profiles 16 crucial conflicts, ancient to modern, that changed the course of Western civilization. Gripping accounts of battles led by Alexander the Great, Joan of Arc, Ulysses S. Grant, other commanders. 27 maps. 352pp. 5⅜ x 8½.   0-486-41129-X

BEETHOVEN'S LETTERS, Ludwig van Beethoven. Edited by Dr. A. C. Kalischer. Features 457 letters to fellow musicians, friends, greats, patrons, and literary men. Reveals musical thoughts, quirks of personality, insights, and daily events. Includes 15 plates. 410pp. 5⅜ x 8½.   0-486-22769-3

BERNICE BOBS HER HAIR AND OTHER STORIES, F. Scott Fitzgerald. This brilliant anthology includes 6 of Fitzgerald's most popular stories: "The Diamond as Big as the Ritz," the title tale, "The Offshore Pirate," "The Ice Palace," "The Jelly Bean," and "May Day." 176pp. 5⅜ x 8½.   0-486-47049-0

BESLER'S BOOK OF FLOWERS AND PLANTS: 73 Full-Color Plates from Hortus Eystettensis, 1613, Basilius Besler. Here is a selection of magnificent plates from the Hortus Eystettensis, which vividly illustrated and identified the plants, flowers, and trees that thrived in the legendary German garden at Eichstätt. 80pp. 8⅜ x 11.
0-486-46005-3

THE BOOK OF KELLS, Edited by Blanche Cirker. Painstakingly reproduced from a rare facsimile edition, this volume contains full-page decorations, portraits, illustrations, plus a sampling of textual leaves with exquisite calligraphy and ornamentation. 32 full-color illustrations. 32pp. 9⅜ x 12¼.   0-486-24345-1

THE BOOK OF THE CROSSBOW: With an Additional Section on Catapults and Other Siege Engines, Ralph Payne-Gallwey. Fascinating study traces history and use of crossbow as military and sporting weapon, from Middle Ages to modern times. Also covers related weapons: balistas, catapults, Turkish bows, more. Over 240 illustrations. 400pp. 7¼ x 10⅛.   0-486-28720-3

THE BUNGALOW BOOK: Floor Plans and Photos of 112 Houses, 1910, Henry L. Wilson. Here are 112 of the most popular and economic blueprints of the early 20th century — plus an illustration or photograph of each completed house. A wonderful time capsule that still offers a wealth of valuable insights. 160pp. 8⅜ x 11.
0-486-45104-6

THE CALL OF THE WILD, Jack London. A classic novel of adventure, drawn from London's own experiences as a Klondike adventurer, relating the story of a heroic dog caught in the brutal life of the Alaska Gold Rush. Note. 64pp. 5³⁄₁₆ x 8¼.
0-486-26472-6

CANDIDE, Voltaire. Edited by Francois-Marie Arouet. One of the world's great satires since its first publication in 1759. Witty, caustic skewering of romance, science, philosophy, religion, government — nearly all human ideals and institutions. 112pp. 5³⁄₁₆ x 8¼.   0-486-26689-3

CELEBRATED IN THEIR TIME: Photographic Portraits from the George Grantham Bain Collection, Edited by Amy Pastan. With an Introduction by Michael Carlebach. Remarkable portrait gallery features 112 rare images of Albert Einstein, Charlie Chaplin, the Wright Brothers, Henry Ford, and other luminaries from the worlds of politics, art, entertainment, and industry. 128pp. 8⅜ x 11.   0-486-46754-6

CHARIOTS FOR APOLLO: The NASA History of Manned Lunar Spacecraft to 1969, Courtney G. Brooks, James M. Grimwood, and Loyd S. Swenson, Jr. This illustrated history by a trio of experts is the definitive reference on the Apollo spacecraft and lunar modules. It traces the vehicles' design, development, and operation in space. More than 100 photographs and illustrations. 576pp. 6¾ x 9¼. 0-486-46756-2